THE TAIG/PEATOL LATHE

and its accessories

Tony Jeffree

Published in the UK by
Tony Jeffree

Copyright © 2019 by Tony Jeffree
All rights reserved.

Except for the quotation of short passages for the purposes of criticism and review, no part of this publication may be reproduced, stored in a retrieval system, or transmitted, in any form or by any means, electronic, mechanical, photocopying, recording or otherwise, without the prior permission of the Author.

Second Edition

This second edition was published in 2018
(The first edition was entitled:
The Taig Lathe and its accessories)

ISBN Number: 9781690196006

Layout and typesetting by
Suidhe Farm Press

Acknowledgments

During the time I have been involved in metalworking, I have received considerable help and advice from other amateur metalworkers, as well as from professional engineers, equipment suppliers and manufacturers. The advent of the Internet has meant that you can now communicate with people in all corners of the globe that have similar interests, but that you are very unlikely ever to meet; to those people, too numerous to mention, that have offered me advice and help, my thanks.

I would particularly like to thank Peter Morrison of Peatol Machine Tools in the UK, Forrest Daley[1] and his family, proprietors of Taig Tools in Phoenix, Nick Carter, a Taig enthusiast turned Taig dealer from Oregon, and Tom Benedict, a Taig enthusiast from the USA, for their help and encouragement.

---

[1] Sadly, Forrest Daley passed away in 2017.

Preface to the Second edition

Sixteen years on since writing the first edition, and various things have changed; I am now retired, I live on a Hebridean island instead of the metropolis of Manchester, and there have been a number of significant developments and enhancements to the Taig (known in the UK as the Peatol) micro lathe. The basic lathe is much the same as it has been for many years, but the addition of an optional power feed system, an optional ER-16 collet spindle headstock, and modification of the cross-slide thrust bearing arrangement are just three of the available upgrades that have made this little lathe even more usable and versatile than before. These enhancements, along with retirement bringing with it a little more free time, meant that it was appropriate to think about a revision of the book to reflect the changes to the lathe and to correct the small number of errors that I have come across in the original text.

Another significant change is that Taig now offer the "Taig Turn" family of lathes based on a modification of the X-Y table from their desktop milling machine - a desktop CNC lathe, a CNC-ready lathe, and a manual lathe; these lathes are significantly more rigid than the original Taig micro lathe, and are equipped with a 5C collet headstock with a through bore that will accept up to 26mm diameter stock. I do not plan to delve into the detail of these lathes in this book, but I will mention some of the accessories for the system that may be of interest to users of the Taig micro lathe.

TONY JEFFREE, 1 SEPTEMBER 2019.

Preface to the First edition

My own introduction to the world of metal machining began in the early 1990s with the purchase of a Taig lathe from Peatol Machine Tools in Birmingham, UK. This came about because of a train of events that started when I was a teenager. I developed an interest in tinkering with clocks and watches and read many books from the library about the subject of clock and watch making. I acquired a small number of rudimentary tools but did not develop my skills beyond very simple repair activities that more often than not didn't greatly improve the performance of the instrument concerned, and sometimes, resulted in the untimely demise of the patient! A major handicap at the time was the lack of access to any serious metal working facilities, or the funds to rectify the situation.

Various other activities, such as University and developing a career in computing and IT, got in the way of any serious attempt to do anything more about clockmaking, until the opportunity to buy a long-case clock for our Victorian house revived my dormant interest in the subject. I again collected a number of books on clockmaking techniques, a few on lathes and metalworking, and decided to buy a small lathe as a prelude to "getting serious" and building my first clock. I settled on the Taig micro-lathe (marketed in the UK as the Peatol micro-lathe), as it seemed to offer good value for money and most of the features that I would need for the purposes of making clock components.

The lathe duly arrived, in kit form; I assembled the lathe and I took my first faltering steps with metal turning. One of the conspicuous gaps in the standard equipment or accessories for the Taig lathe at that time was that there was no kind of manual available that would give the novice metal turner any clue as to where to start, either with proper assembly and adjustment of the lathe, or with its subsequent use. There were a couple of sheets of photocopied notes describing how to lap the cross slide to the bed, but that was about all. The absence of such a manual is a major part of the motivation for this book. There may be material in the book that the more experienced Taig user will already be familiar with, and which can therefore be skipped over. However, I have included material that describes the construction of various accessories for use with the lathe, which should be of interest to the more experienced user, while still being within the machining capabilities of the relative novice.

It was not my intent to produce a book that deals with basic workshop practice and machining techniques; there are many other good books available that will serve that purpose better than this book can. However, my hope is that the material between these covers will at least help someone new to machining to get started with their new Taig lathe or mill.

Metalworking often involves considerable expenditure of time and effort aimed at making the tools, fittings and fixtures that are required in order to machine the components that you actually wanted to build in the first place. For example, if your intention is to cut a gear wheel for a clock mechanism, your first problem is to devise a way of accurately dividing a circle into the number of divisions required to cut the gear teeth; so, before cutting the gear wheel, you might find yourself diverted into building dividing accessories, or maybe a dedicated "wheel-cutting engine", in order to carry out the original task of cutting the gear wheel. Building the dividing head, or wheel-cutting engine, or whatever, will inevitably involve making further fixtures and fittings to carry out the particular machining sequences involved...and so on.

For some people, making these tools, fixtures and fittings can become an end in itself, as there is considerable satisfaction to be obtained from making a well-designed, well-finished, functional tool. For others, this is just an inevitable, sometimes frustrating, part of the process, and the final finished object (the clock, or steam engine, or whatever) remains the final goal. I am somewhere in the middle of these two positions; I derive considerable pleasure from making tools and fixtures, but also from making things that would be more recognizable to the uninitiated as functional or simply ornamental. Twenty years or so on from my initial lathe purchase, I have actually succeeded in building a wall clock that tells the time with a fair degree of accuracy, and which looks extremely decorative in our lounge, in its brass-framed glass case; all of the turning and milling operations involved in its construction were performed on my Taig lathe and Taig CNC mill. In the process of developing the tools needed to build the clock, I have had a lot of fun building accessories and improving the versatility of the lathe, and some of these accessories and improvements appear in the following pages.

TONY JEFFREE, 16 JANUARY 2003.

Table of Contents

The Taig/Peatol Lathe 1
Acknowledgments 1
Preface to the Second edition 2
Preface to the First edition 3
Chapter One 9
    Applications 10
    Safety 11
Chapter Two 14
    Assembling the lathe 14
    Choosing a motor 17
    Mounting the lathe 19
    Motor mounting techniques 21
    Maintenance 24
Chapter Three 25
    The 3-jaw chuck 25
    The 4-jaw independent chuck 27
    The 4-jaw self-centering chuck 28
    The chuck adapter 29
    The face plate 30
    Tailstock drill chucks 30
    Drill chuck arbor 31
    Taig collet set 32
    WW collet headstock 33
    ER16 collet headstock 33
    5C collet headstock 34
    ER16 collet adapter 35
    Blank and special purpose arbors 35
    Tailstock support 37
Chapter Four 39
    Tool posts 39
    HSS tool bits 42
    Carbide tools 43
    Replaceable tip tools 43
    Sharpening lathe tools 45
    Boring bars 47

Making a boring bar holder..48
Radius/ball turning tool 51
Slitting saw...52
Using cutting fluid..52

Chapter Five..54

The lever operated drilling tailstock 54
Needle bearing centre..55
Milling slide and milling vice...55
Milling cutters 56
Compound slide 57
Steady rest..58
Riser blocks 59
Chuck depth stop..59
Grinding wheel set 60
Tailstock die holder..60
Spindle wrench..61
Wood-turner's face plate 62
Adjustable tool rest 62
Lathe dog..63

Chapter Six..64

Care of the lathe 64
Fitting and removing chucks 66
Using collets...67
Selecting the right cutting speed 67
Parallel turning 69
Turning between centres 70
Taper turning 72
Milling...72
Power feed 76

Chapter Seven...78

Spindle runout 78
Axial alignment...79
Checking axial alignment quickly 81
Checking axial alignment accurately..82
Adjusting the tailstock - quickly 84
Checking tailstock alignment - accurately..................................86

Chapter Eight..88

    Spinning handles 88
    Locking knobs and handles 89
    Extending the tailstock lever 91
    Micrometer set-over adjuster 92
    A filing rest 95
    Machining small screws 98
    Handwheel adapter 101
    Additional chucks 103
    A tailstock travel indicator 105

Chapter Nine 107
    Simple dividing operations 108
    Graduating using the lathe 110
    Dividing heads and rotary tables 110
    Headstock dividing attachment 112
    Building the dividing attachment 115
    Modifying the headstock 116
    The brake shoe 120
    The pinch bolt and tommy bar 122
    The dividing assembly mounting plate 124
    Modifying the 20dp worm wheel 126
    The division components 126
    The primary worm carrier 127
    The secondary worm assembly 128
    The primary worm shaft 132
    Index plate boss 132
    The sector arms 134
    Division plates and hole circles 136
    The indexing arm components 138
    Tailstock support 141
    Putting it all together 141
    Direct dividing setup 141
    Simple and compound dividing setup 142
    Simple division 143
    Compound division 143
    Using a protractor 144

Chapter Ten 145
    Attaching the drive to the spindle 146
    The tumbler support bracket 149

The tumbler assembly	151
The dog clutch	158
Fitting the clutch	164
The leadscrew	166
The split nut assembly	171
The "banjo" components	176
Fine feed set-up	180
Screw cutting set-up	182
A word on safety	184
Annex A	**185**
Taig Tools	185
Taig dealers	185
Tool suppliers	185
Other suppliers	186
Internet resources	186
Annex B	**188**
General Specifications	188
Capacity	189
Spindle	189

Chapter One
Introduction

The Taig lathe (marketed in the UK as the Peatol lathe) has developed a well-deserved reputation and a committed following during the 40 years or so that it has been in production. Designed by Forrest Daley, an engineer who spent 18 years working in the aerospace industry, the lathe is a fine example of the application of sound engineering principles, and of the use of appropriate materials and production methods to produce a cost-effective end result. More recently, the Taig lathe was joined by the Taig mill, visibly from the same stable, as it uses the same headstock as the lathe and, like the lathe, makes use of a combination of steel and extruded aluminium components in its manufacture.

Taig lathes and mills are built entirely in the USA by Taig Tools, in factory premises housed in the grounds of the Daley home near Phoenix, Arizona. The factory itself is a good example of pragmatic engineering practice and ingenuity. Taig Tools still make use of old technology where appropriate; "automatic" lathes that munch their way unattended through long lengths of bar-stock, for example, and which have by now paid for themselves many times over. At the other extreme, pallet-loaded CNC machining centres are also used to good effect. The racks used on the Taig lathe are machined using screw cutting techniques; lengths of steel bar stock are loaded into longitudinal slots in a large metal drum, and the whole assembly is mounted in a screw cutting lathe set to the right screw pitch to cut the rack teeth. The large diameter of the drum means that the inevitable helix angle of the rack teeth is small enough to ignore.

At the time I wrote the first edition, something like 40,000 Taig lathes had been sold; no doubt in the intervening years that number has increased considerably. This is a reflection both on the utility of this versatile machine, and on the fact that it is highly affordable. The Taig lathe has some strong competitors in the "desktop machining" world, some that may be cosmetically more appealing, others that may appear to be more versatile or cheaper. The fact that it is still competitive in versatility, performance, and price after more than 40 years in production speaks for itself.

Wherever I have remembered to do so, I have included Taig part numbers in braces; for example (#1160) gives the part number of the 20" drive belt. The reader can then quickly relate a component or accessory mentioned in the text to the corresponding item in the Taig price list.

Applications

The Taig lathe is one of the class of lathes that are used for "desktop machining". It is small enough that it can be attached, along with its drive motor, to a base-board, used on a workbench or desk, and then packed away in a cupboard when not in use. Its capacity is not large; it can swing a work piece that is up to 4.5" in diameter and up to about 9" between centres. However, at this extreme of its operation, attempting to machine hard materials like cast iron or steel is a challenge that has to be approached with very sharp tools and very light cuts.

It is possible to use "riser blocks" to increase the spindle height of the lathe by an extra inch; however, this reduces the stiffness of the machine, and is therefore only of use with softer materials and extra care.

The major applications for this kind of lathe tend to be at the small end of what is often known in the UK as "Model Engineering" and in the US as "Home Shop Machining"; making scale models of various kinds, clockmaking, small steam engines...and so on. However, they are also widely used in small commercial production and prototyping activities.

It is a mistake to view these lathes as only metalworking lathes. There are many users that have bought Taig lathes for machining other materials, such as plastics and wood for pen-making. There are even users of somewhat modified Taig lathes that use them for machining pool cues.

In my own association with these machines, the only times that I have felt the Taig lathe to be at all limited have been when I have needed to machine stock that is physically beyond the lathe's capacity; as my main preoccupation in machining has been making clock parts, this hasn't happened very often. As with any machine tool, the Taig lathe has a limited operating envelope in terms of depth of cut and rate of cutting ("speeds and feeds") that it can support; this envelope will depend on the diameter and type of material being machined, the work holding method employed, and the type of cutting tools in use.

There is no point in pushing any machine beyond its operating envelope and expecting it to perform well. The only possible end results of such abuse are damage to the machine or cutting tools, poor surface finish, destruction of the component, and frustration on the part of the operator.

Safety

The Taig lathe is a serious metal cutting tool; as such, it is quite capable of damaging or destroying body parts should they come into contact with it in inappropriate ways. Therefore, as with all potentially dangerous machinery, it should be treated with respect and used with proper care and preparation. The following are some tips (and this MUST NOT be regarded as a comprehensive list) that may help you avoid your association with your lathe becoming an unpleasant or disabling experience!

— Never allow body parts, hair, or loose clothing to come into contact with the rotating parts of your lathe. If you have long hair, tie it back out of the way. If you are ever tempted to wear loose clothing near the lathe, DON'T!

— Fit guards to cover the "back end" of the machine (rear of the spindle, drive belt and motor shaft), and the chuck/spindle. The former will make it harder to get tangled up in the drive pulleys; the latter will make it harder to trap fingers in the chuck and may also help contain fast-moving swarf particles and prevent them embedding in the operator's skin. When taking heavy cuts in metal, the particles of swarf can be red hot as they leave the workpiece.

— Be aware of the properties of the materials that you are machining, and the potential hazards that they represent. Titanium and Magnesium are examples of flammable metals; it is possible for the heat generated during cutting to ignite particles of Titanium or Magnesium swarf as they leave the workpiece, which can in turn ignite other materials in the workshop (including piles of Titanium and Magnesium swarf from previous machining operations). Other finely divided metals can also be flammable; for example, steel wool can be easily ignited with a match. Fires of this kind are often very difficult to extinguish. If you plan to work with materials that represent a fire hazard, consult with your local fire department as to suitable precautions and types of extinguisher that are appropriate for your use, and, as a general rule, clean up and dispose of swarf before it can become a hazard.

— Do not store flammable swarf (including finely divided metals) in waste bins in the workshop.

- Use proper protective gear for the job in hand. In particular, use safety glasses or goggles AT ALL TIMES to guard against flying particles.
- Never allow children to use the lathe or allow anyone to approach within the "danger zone" while you are using it. Bear in mind that the "danger zone" includes any clear space you may urgently need to occupy to get away from a piece of machinery that, despite all your good intentions, is getting out of control.
- If necessary for the safety of others, keep your machinery locked away when you are not using it.
- Assembling a Taig lathe involves some minor electrical work, to wire up the motor and a suitable on/off switch. If you are not competent to do this work (and please note, the word was *competent*, not *confident*!), find a qualified electrician to do it for you.
- When you use a chuck or a face plate, make certain that it has been firmly tightened onto the spindle nose before use. Vibration during cutting can cause a chuck to loosen; when you switch off the motor, the chuck or faceplate can then unscrew and fall off the end of the spindle. Depending upon the spindle speed in use, this can be a rather more exciting experience than you really need. A 4-jaw chuck, spinning at 1000 RPM, complete with work piece, falling onto the lathe-bed and then launching itself across the workshop, is a frightening spectacle that can result in real damage both to the lathe bed and to the operator.
- Always make sure that your work piece is properly secured to the face plate or chuck, and check that it can be rotated without fouling the lathe bed or carriage, before you start machining. Give the spindle a full 360-degree rotation by hand before switching on.
- Never leave chuck keys in the chuck when you have finished using them. These turn into projectiles if not removed before the lathe is switched on!
- Never use machinery if you have consumed alcohol or any other drug, prescription or otherwise, that may affect your judgment or speed of reaction, or if your performance is likely to be affected by lack of sleep.
- Make sure that the lathe cannot be accidentally started when it is plugged in. The ideal is to use a "no-volt release" type of power switch as the on-off switch for the lathe, and preferably one with an "emergency stop" button.
- Make sure that the "OFF" switch can be reached quickly from the front of the lathe, without having to reach across the working area.
- When machining long, thin stock in the lathe, make sure that both ends are properly supported. Failure to do this can result in free ends "whipping" at

high rotational speeds, with corresponding danger to the operator or spectators.
— If you use the grinding accessories with the lathe, take care to protect the lathe bed and cross slide from dust. The abrasive particles shed from grinding wheels do not mix at all well with the lathe's bearing surfaces. Preferably, strip the lathe down and clean it thoroughly after this kind of use. Failure to do this will severely affect the lathe's working life.
— Keep your lathe tools sharp, and with the right cutting angles and clearances for the materials being machined. Blunt or badly formed tools can lead to excessive cutting forces and potential damage to the machine, quite apart from the fact that they will produce a poor surface finish.
— Don't try to force the lathe beyond its capabilities. This will only lead to frustration, and to potential damage to you and the lathe.
— It is worth bearing in mind that while it is always possible to replace broken lathe components and work pieces, it may be impossible to replace damaged body parts. Hence, if things start going badly wrong, it is better to get out of the way and switch off the power until the dust has settled, rather than attempt to retrieve what may be a dangerous situation that cannot be salvaged.
— Don't be tempted to work at the lathe while sitting down, even if it seems more comfortable that way; mount the lathe at a height that is comfortable for working while standing up. If a lathe chuck comes loose and unscrews from the spindle (see earlier), it is much easier to get out of the way if you don't have to wrestle with a chair at the same time.

Chapter Two
Assembly and mounting

The Taig lathe is unusual in that it can be purchased in "kit" form; the basic lathe has no motor or base-board, and depending upon which version is purchased, the lathe itself may require assembly and "lapping" of its sliding components before it is ready for use. This helps to keep the cost of the lathe competitive, at the expense of a bit of additional work before the lathe can be used; however, it could be argued that if you're not comfortable with assembling a lathe of this kind, then perhaps metalworking is not for you!

The following sections describe the assembly and setup of the lathe and suggest a few possible solutions to the problem of powering the lathe.

Assembling the lathe

If one of the "Factory assembled Micro Lathe" variants (#L1015, #L1017, M1015,...etc) has been purchased, then most of this section can be skipped; however, the assembly instructions below include some comments on adjusting the cross slide and carriage gibs.

If the "Unassembled Micro Lathe Kit" (#K1019) has been purchased, the first step will be to assemble the lathe components. Referring to the exploded diagram of the lathe[2] (Figure 1):

---

[2] These instructions refer to the current style of lathe; there are some detailed differences, and differences in part numbering, for the earlier versions of the lathe.

THE TAIG/PEATOL LATHE

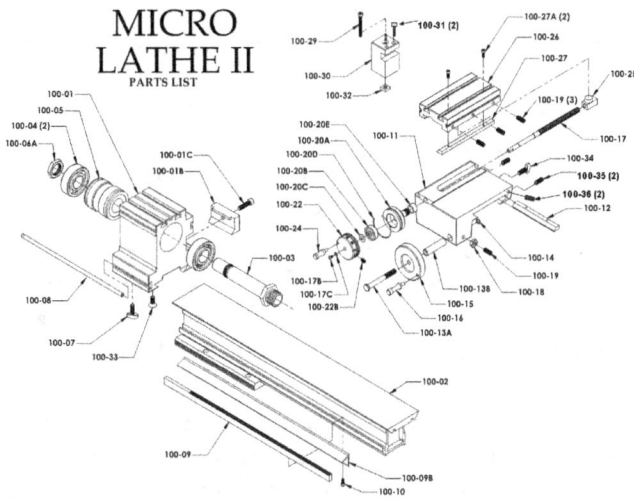

Figure 1 - Exploded diagram of the Taig lathe

— Install two socket head set screws (#100-35, size 10-32 UNF) in the outer pair of holes in the rear face of the carriage (#100-11). These set screws are used to adjust the fit of the gib strip (#100-12). A thumb screw (#100-34) is fitted in the hole between these two set screws; the thumb screw is used to lock the carriage if necessary.
— Fit the gib strip in place and slide the carriage onto the end of the lathe bed. The two set screws should be tightened to indent the back of the gib strip, then loosened to a snug, sliding fit. Indenting the gib strip prevents the strip from sliding independently from the carriage.
— At this point, the carriage needs to be "lapped" to the bed, to ensure that the carriage bearing surfaces are properly bedded in, and therefore, that the whole of the bearing surface is being used to support the carriage. This step helps to ensure the longevity of the lathe. A lapping paste can be made up by mixing a powdered, mildly abrasive, household cleanser such as Comet (USA) or Ajax (UK) with oil to form a thin paste. It is important to choose an abrasive that is "non-embedding"; many of the harder abrasives (diamond, silicon carbide, aluminium oxide, etc.) are capable of embedding particles in soft metal surfaces, resulting in accelerated wear of the sliding surfaces of the lathe.
— Apply the abrasive paste to the bearing surfaces and slide the carriage back and forth, adjusting the gib strip and applying more paste every few strokes.

The lapping process should be complete after 30-40 strokes; then remove the carriage and wipe off all traces of lapping compound from all components, oil and re-assemble.

— The cross slide (#100-26) is then assembled. The cross slide gib (#100-27) is held in place by two small socket head screws (#100-27a) through the top of the slide, and three set screws (#100-19) provide adjustment of the gib. The middle screw is adjusted first, then the two end screws.

— Lap the cross slide to the carriage using the same technique as before, again, cleaning off all residue and re-lubricating prior to reassembly. Insert the cross nut (#100-25) into the hole in the underside of the slide; this nut engages with the cross slide's leadscrew to give controlled movement of the cross slide.

— The cross slide leadscrew assembly is supplied pre-assembled and adjusted. Fit the assembly to the carriage by screwing the threaded end of the bearing block into the threaded hole at the front of the carriage. A drop of Locktite on the threads will help ensure that the assembly doesn't come loose in use.

— Slide the cross slide carefully into place, engaging the leadscrew with its nut in the process. The gibs should be adjusted to remove all play in the cross slide while still allowing easy movement when turning the handwheel.

— Install the rack (#100-09) in the channel provided on the side of the lathe bed. The rack is held in place by a single socket-head screw (#100-10, size 4-40 UNF), at the right-hand end of the bed.

— The carriage handwheel assembly consists of the carriage handwheel and shaft (#100-15, #100-16, and #100-13a), the eccentric bushing (#100-13b), and the retaining circlip (#100-14). The eccentric bushing is pushed into the hole at the bottom of the carriage apron, the handwheel shaft fitted through the bushing, and the circlip is fitted to the end of the shaft to retain the shaft in the bushing with minimal end-float. The bushing is rotated to adjust the play between the pinion gear at the end of the shaft and the rack, and a set screw and locknut (#100-19, #100-18) are used to lock the bushing in place. The carriage should now be able to traverse from one end of the lathe bed to the other, under control of the carriage handwheel.

— The headstock is pre-assembled; this can be slid onto the dovetail of the lathe bed at the left-hand end (the end of the bed with the mounting "foot"), with the threaded spindle nose pointing to the right. A socket head screw (#100-01C) passes through the clamping piece at the back of the headstock; this should be tightened in order for the headstock to grip the lathe bed. Do not over-tighten the screw; overzealous application of the hex

wrench is unnecessary and can distort the headstock, resulting in possible misalignment of the spindle with the bed.
— The tool post (#100-30) is fitted to one of the T-slots on the top of the cross slide, using the T-bolt and nut (#100-29 and #100-32). The two set screws (#100-31) are used to clamp a suitable cutting tool in the tool post's slot.
— Finally, the saddle stop rod (#100-08) is fitted into the channel in the foot of the headstock, and a thumbscrew (#100-07) fitted under the foot to lock the saddle stop in the desired position.

At this point, you should have an assembled lathe unit that should look remarkably similar to the one in Figure 2, perhaps minus the tailstock and 4-jaw chuck.

Figure 2 - An assembled Taig lathe

Choosing a motor

As the lathe does not come with a pre-assembled drive unit, the user is free to choose from a wide range of possible drive units.

The lathe does not require a great deal of power; 1/4 hp is about right. As the lathe uses a pulley system to drive the spindle, using a more powerful motor than this is not a problem if you have one to hand, as the belt will slip if you attempt to make the lathe do too much work. Even with a 1/4 hp motor, it is not possible to deliver all the motor's potential output through the lathe pulleys.

Broadly speaking, the options for driving the lathe are as follows:
— Use a conventional fixed speed, single phase, capacitor start motor running off the mains (110v in the US, 220 or 240v in Europe). Taig in the USA and Peatol in Europe offer suitable motors of this type.

- Use a 3-phase, fixed speed motor, driven from a 3-phase supply or a phase convertor.
- Use a 3-phase motor, driven from a "Variable Frequency Drive" (VFD), giving variable speed control.
- Use a DC motor, running from a suitable variable speed control. There are a wide variety of such units available, including so-called "treadmill" motors that are supplied by surplus dealers, and sewing machine motors/controllers.

Of these options, the cheapest and simplest will probably be a conventional fixed speed, capacitor start motor, or a 3-phase, fixed speed motor if you have a 3-phase supply readily to hand.

Table 1 - Spindle speeds

Motor Pulley	1750 RPM Motor	1400 RPM motor
(smallest) 1	525	435
2	825	685
3	1300	1085
4	2100	1750
5	3350	2790
(largest) 6	5300	4415

Variable speed is a luxury that, once you have had it, you will be reluctant to give up, but for most purposes, it is not really necessary. However, if you can obtain a variable speed motor at a reasonable price, go for it. It is also possible to obtain 2-speed capacitor start and 3-phase motors; this can sometimes be a useful compromise. If you can afford to do it, a 3-phase motor with a VFD is a great solution, as it is possible to drive the motor with relatively constant torque over a wide range of speeds. A VFD delivers variable frequency 3-phase current to the motor; I have such a drive on my Myford lathe, where it delivers 3-phase that can be varied from 5 cycles/sec through to 120 cycles/sec, and hence delivers a motor speed range from 10% of its nominal RPM, through to 240% of its nominal RPM. With this kind of speed range, it is possible to keep the lathe on one pulley setting for most operations.

Figure 3 - The Taig 6-step pulley set

As the lathe in its basic form has no leadscrew, there is no pressing need to have a motor that can be reversed, although a reversible motor can come in handy if you are likely do a lot of threading work in the lathe with taps and dies.

Capacitor start motors generally deliver 1725 RPM on a 60-cycle mains supply, or 1400 RPM on a 50-cycle supply. Table 1 gives the approximate RPM values that can be expected when using these motors with the standard Taig pulley set, which is shown in Figure 3. If a variable speed solution is chosen, it is worth checking what RPM range the motor will give, and hence, what range of spindle speeds will result, to make sure that this will be suitable for your needs.

Also, bear in mind that the spindle bearings are rated for a maximum of 7000 RPM, so attempting to exceed this speed is not advisable, nor, for most purposes, is it necessary.

Mounting the lathe

The next consideration is a suitable mounting arrangement for the lathe. To some extent, the choice of mounting will depend on the uses to which the lathe will be put, and to the choice of motor.

If you want a fixed installation and have a handy bench, the lathe and motor can be mounted directly on the surface of the bench; as the lathe bed has a cantilevered mounting arrangement (it has a mounting foot at the headstock end, and the tailstock end of the bed is free), it is not necessary to mount the lathe on a metal surface or to use the adjustable mounts that are used to "level" the beds of larger lathes.

If a movable installation is desirable (for example, so that the lathe can be stored out of the way), then the lathe can be mounted on a base-board, which can be made of metal plate, timber, particle board, etc., depending upon what is available.

A bit of thought needs to be applied to the positioning of the motor relative to the lathe, and you will need to check which way the motor rotates in order to determine where to mount it.

Figure 4 - Example of a baseboard layout

The lathe spindle should rotate counter-clockwise when the spindle is viewed from the tailstock end of the lathe; in other words, it rotates in the direction you would expect a power tool to rotate if you were drilling a hole with a conventional drill bit. So, if the spindle of the motor you have chosen rotates clockwise when viewed from the business end of the spindle, looking along the spindle's axis, then the motor will need to be mounted behind the lathe, with the spindle projecting to the left; this arrangement can be seen in Figure 4, where both the lathe and the motor have been mounted on a timber baseboard. If the motor rotates counter-clockwise, then you will need to mount the motor to the left of the lathe, with the spindle pointing to the right; clearly, this will result in a longer "footprint" for the baseboard, as the motor will extend to the left of the lathe.

It is worth considering where you will place the on/off switch when choosing a baseboard layout; the switch needs to be accessible without reaching across the machine, and without bringing your hands too close to moving parts. My own preference is to mount the switch to the right of the lathe bed, but others may prefer to mount it in front of the lathe at the headstock end. Either way, make sure that the chosen position doesn't interfere with the operation of the lathe, or with removal of the saddle from the bed for cleaning and maintenance purposes.

Motor mounting techniques

As a belt drive system is used with the Taig lathe, it is necessary to mount the motor in such a way that the belt tension can be adjusted quickly and easily. This is particularly important for changing lathe speeds, as this is achieved by moving the belt from one pair of pulleys to another; attempting to change speeds without removing the belt tension first will result in damage to the belts.

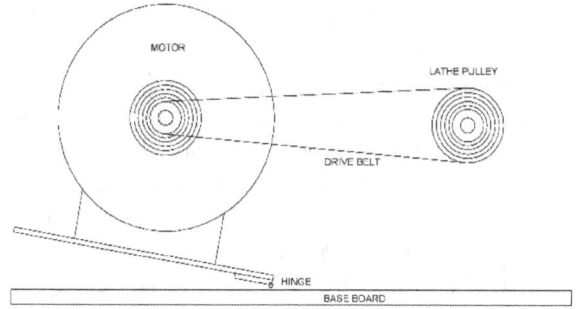

Figure 5 - Simple hinged motor mount

The simplest approach is to mount the motor so that it is hinged at the edge nearest to the lathe and use the weight of the motor to tension the belt. This approach is illustrated in Figure 5; if the weight of the motor is such that this type of mount puts too much tension on the belt, a simple screwed stop could be added to the motor foot to adjust the belt tension, while still retaining the simplicity of lifting the motor up to change pulleys. Alternatively, if the motor has a tendency to bounce under load, more weight could be added to the motor board or the motor moved further away from the hinge.

Figure 6 - Peatol motor and mounting plate

Peatol Machine Tools in the UK sold a mounting plate assembly that accommodates the lathe and their standard capacitor start motor, as shown in Figure

6. The motor mounts on a raised plate, hinged at the back by two resilient rubber bushes, and provision for adjusting belt tension is provided by the adjustable bolt at the front of the motor plate (the mounting plate is shown in rear view in the photo). The lathe is bolted to the base plate in the space next to the adjuster. The mounting plate has to be drilled for the motor and lathe foot mounting bolts before it can be used.

Taig supply drive belts in two sizes; either 12 1/2" (#1159) or 20" flat length (#1160). The longer belt is needed if the motor is to be mounted behind the lathe (as in Figure 4), in order to allow sufficient space between the motor and the lathe. If the motor is mounted to the left of the lathe, then either belt is potentially usable.

The motor mounting arrangement used on the Taig mill, which uses the same headstock as the lathe, is adaptable for use with the lathe in cases where the motor is mounted to the left of the headstock. This approach is illustrated in Figure 7, which shows a plan view of the lathe headstock and motor mount. A 5" length of 1" square aluminium bar is attached to the rear T-slot at the foot of the headstock, by means of two 10-32 UNF bolts and square nuts, and a square aluminium plate is attached to the end of the bar by means of two 10-32 UNF bolts trough the corner of the plate. The bolt labelled A passes through a plain hole in the plate; the second bolt passes through a curved slot (radius equal to the separation of the two bolts), which allows the plate to pivot around bolt A. A hole through the centre of the plate for the motor spindle allows the front of the motor to be bolted to the plate.

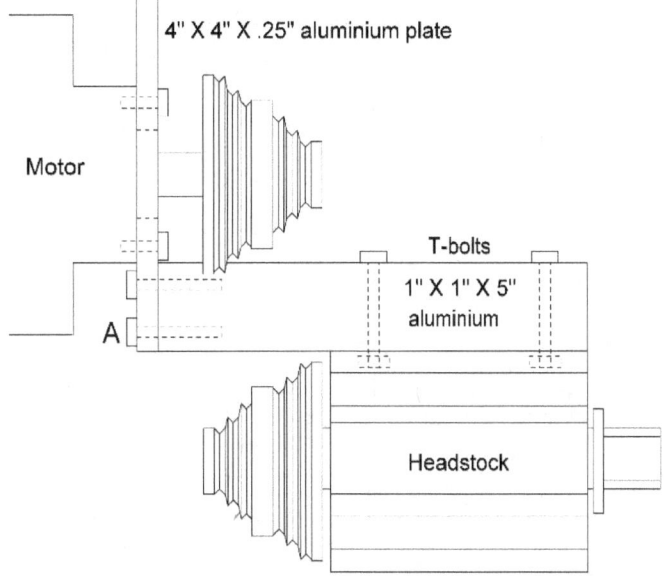

Figure 7 - Mounting a motor to the left of the headstock

Adjusting the tension in the belt is a matter of slackening the two bolts holding the plate to the bar and rotating the motor towards or away from the lathe. The two T-bolts holding the bar to the headstock allow the position of the motor pulley to be adjusted relative to the headstock pulley, to ensure that the two pulleys are perfectly aligned. This mounting arrangement is suitable for use with the shorter of the two belts.

Regardless of which motor mounting arrangement is used, proper alignment of the pulleys is very important; failure to do this will severely reduce the life of the drive belts. The simplest way to achieve correct pulley alignment is to apply a straight edge across the parallel sections of the two pulleys.

A worthwhile tip before fitting the pulley to the lathe spindle is to file a small flat on the spindle to take the pulley's set screw. The lathe pulley is a very close fit, and failure to file a flat will make the pulley very difficult to remove, as the set screw will raise a burr on the spindle that will lock the pulley in place.

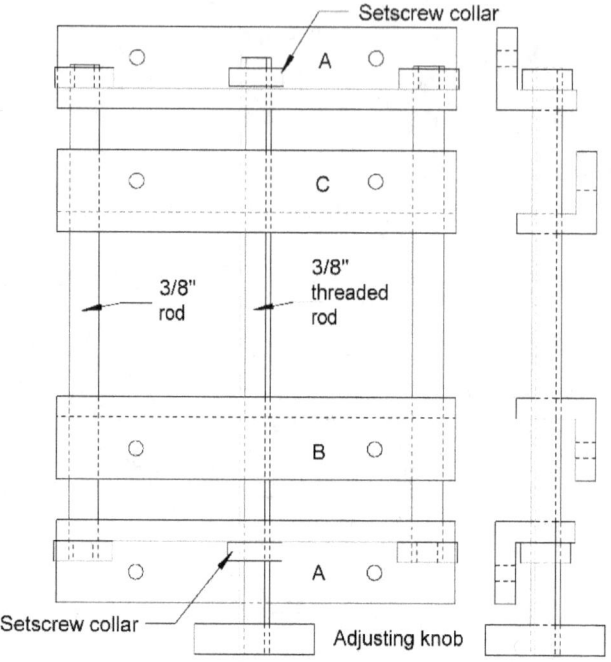

Figure 8 - Sliding motor mount

More complex motor mounting arrangements are possible for the more ambitious constructor; for example, Nick Carter, a Taig dealer from Oregon, favours a motor mount fabricated from four 1/4" thick aluminium angle brackets, a pair of 3/8" steel rods, and a 3/8" leadscrew, allowing the motor to slide back and forth under the control of the screw. A rough sketch of this mount is shown in Figure 8. The pair of brackets marked "A" are screwed to the base-board; the brackets marked "B" and "C" carry the motor, and therefore, the holes drilled for the motor need to be positioned according to the spacing of the motor's mounting bracket. The two plain rails, both of 3/8" steel rod, are threaded at either end so that they can be secured in place with suitable nuts. The leadscrew is held in place with setscrew collars at either end, so that it can rotate freely in the two A brackets. Brackets B and C should be drilled to allow a sliding fit over the two rails; bracket C is drilled for a sliding fit over the leadscrew, but bracket B is tapped to match the leadscrew thread.

All four brackets should be clamped together when drilling the holes for the rails and leadscrew, to ensure that the hole spacing is exactly the same on all brackets; otherwise, it is unlikely that the motor brackets will slide satisfactorily on the rails.

The periphery of the knob is knurled to give a good grip; the knob is tapped to match the leadscrew thread and held in place with a drop of Locktite on the threads.

Maintenance

The Taig lathe requires very little maintenance; however, the lathe should be kept clean. This is particularly important when the lathe has been used with abrasive materials; for example, when machining cast iron or using the grinding wheel attachments. Surprisingly, dust from machining some types of wood can also be very abrasive and should be carefully removed.

After working on the lathe, clean all debris from the lathe with a small brush and a soft cloth. I find a 1" paintbrush is useful for this purpose. Thoroughly clean the spindle threads, and the corresponding threads in the backplates of the chuck or face plate, to remove any swarf; this is particularly important when machining aluminium and other soft metals, as particles can become embedded in the threads, making it progressively more difficult to attach and remove the chucks and face plates.

When the lathe has been cleaned, oil the bed and the cross slide ways with a light machine oil. A light smear of oil on any steel components will help to prevent surface rust developing.

If the lathe is to be stored unused for long periods, remove any tools from the spindle to make sure that they cannot rust in place as a result of damp penetration.

The spindle bearings themselves are "lubricated for life" and should not need additional lubrication.

Chapter Three
Work-holding devices

There are a variety of work-holding devices available for use with the Taig lathe, each suited to a particular type of use. Understanding the uses and limitations of these devices is an important first step in using the lathe.

The 3-jaw chuck

Often known as a self-centering or "scroll" chuck, the chuck jaws move in and out simultaneously as the scroll is rotated in either direction. This chuck is primarily used for holding round or hexagonal section stock; as it has only three jaws, it cannot be used to hold square section material.

This is one of the simplest work-holding devices to use, as it will hold round section stock roughly on centre and can be operated very quickly. However, as with most self-centering chucks, it cannot be relied upon to hold stock accurately on centre, as the scroll mechanism is very rarely sufficiently accurate for the centering to be "spot on". Hence, if it is necessary or acceptable to machine the entire outer surface of the part, then using the 3-jaw chuck may be appropriate; however, if you wish to hold a previously machined, round section part accurately centered, and perform a second, concentric machining operation on it, then this is not the chuck to use.

The Taig 3-jaw chuck (#1050) can be seen in Figure 9. An unusual feature of this chuck is that it has replaceable soft aluminium jaws. This makes it possible to purchase additional jaw sets (#1051) that can be machined for particular purposes; for example, steps can be cut in the outer edges of these jaws to allow the chuck to grip inside tubular items, or internal steps cut to allow large diameter items to be gripped.

Figure 9 - The 3-jaw chuck

Before using the chuck, the inner faces of the jaws need to be trued. This is achieved by opening the jaws wide enough to accept a washer of about 1" diameter; the washer is held as far back in the jaws as it will go, and the scroll tightened using the tommy bar provided with the chuck.

A UK 10-pence piece or a US quarter is about the right size for this. A boring bar (see "Boring bars" on page 37) is fitted to the tool holder, and the inner faces of the jaws are machined until all three are "cleaned up". This serves two purposes: firstly, to make sure that the gripping faces of the jaws are parallel to the lathe spindle's axis, and secondly, to make sure the jaws grip as nearly on axis as possible. Closing the jaws on the washer is important in order to prevent the jaws unscrewing during the machining, and also to ensure that the jaws are held as they would be when gripping a real workpiece.

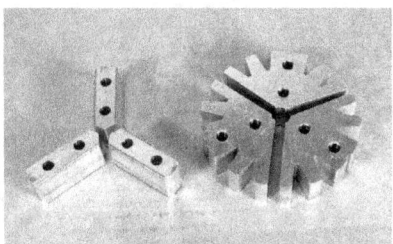

Figure 10 - Replacement soft jaw sets

This machining will leave a small pip at the back of each jaw where the washer was gripped; this must be removed by careful use of a small metal file.

This jaw trueing procedure works best when the jaws are closed on a washer that is close to the diameter of the workpiece that will be gripped by the jaws. If it is desirable to machine the jaws so that they are very accurately on-centre for a particular workpiece, it can sometimes be helpful to close the jaws completely, and then bore the jaws to slightly less than the workpiece diameter, although this will work only for small diameters due to the position of the mounting screws for the jaws.

In addition to the standard jaw sets, Taig supply "full circle" jaw sets (#1052) that can be used with this chuck; each jaw is a 120-degree sector of aluminium (Figure 10). These are intended to be used to hold thin-walled tubing or other fragile components without distortion; again, machining suitable steps in these jaws allows them to be adapted to the size of tube to be gripped and can be used to grip internally or externally.

As the jaws used with the 3-jaw chuck are simply pieces of aluminium bar drilled for the two mounting bolts, it is also possible to make your own jaws for special purposes if need be, using whatever material is appropriate to the job in hand (steel, brass, aluminium...etc.).

A version of this chuck is available () that screws onto the spindle nose of the ER16 headstock (#1050ER, see *ER16 collet headstock* on page 33).

The 4-jaw independent chuck

The 4-jaw independent chuck has, as the name implies, jaws that are adjusted independently. These chucks are used to hold round, square or irregularly shaped work pieces, and allow the work piece to be accurately centered if this is necessary for the machining operation involved.

Figure 11 - The 4-jaw independent chuck

The Taig 4-jaw independent chuck (#1030) is shown in Figure 11. The jaws are of hardened steel and are reversible, allowing a wide range of shapes and diameters of work piece to be held.

Adjustment of the position of each jaw is achieved by means of a hexagonal "Allen key" wrench. This independent jaw adjustment means that it is possible to accurately centre round stock in the chuck, with the aid of a dial indicator. The work piece is gripped in the chuck roughly on-centre, and the dial indicator set up so that its plunger is in contact with the outside surface of the work piece. The various T-slots on the lathe headstock and tailstock can come in handy for mounting dial indicators; alternatively, a stand with a magnetic base can be used to mount the dial indicator on the lathe bed.

The chuck is rotated slowly by hand; the dial indicator reading is used to detect the maximum and minimum deviation, the difference between these being twice the amount by which the part is off-centre.

More usefully, the dial indicator allows you to determine how much to move a given pair of jaws to remove their contribution to the error. Note the reading when a given jaw passes the point of contact with the dial indicator and take a second reading when its opposite jaw passes the dial indicator point; the difference is twice the error contributed by that pair of jaws.

Trial and error will give you a feel for how far the adjusting screws of the jaws have to be turned to remove a given error; as the screw pitch is quite coarse, a small movement of the screw has quite a large effect. However, with careful use, it is possible to use this chuck to centre a work piece to a high degree of accuracy, allowing second operations to be performed with a reasonable degree of concentricity.

Obviously, the 4-jaw chuck is not as quick and easy to use in cases where accurate centering is not important, and this explains the relative popularity of the 3-jaw chuck. However, given the versatility of the 4-jaw, this is the chuck of choice if your funds do not stretch to buying both.

A version of this chuck is available that screws onto the spindle nose of the ER16 headstock (#1030ER).

The 4-jaw self-centering chuck

The 4-jaw self-centering chuck (#1060) has jaws that are adjusted by a scroll, in the same way as the three-jaw self-centering chuck (#1050). These chucks were primarily aimed at woodworking uses; unlike the 3-jaw chuck, which can hold round stock securely, because the two opposing pairs of jaws may not close at exactly the same rate, the 4-jaw self-centering chuck is likely to grip round stock (or square stock, for that matter) firmly with only two of its four jaws. This need not be a problem if you are gripping material that has some "give" such as wood or plastic, in which case additional torque on the scroll will give good grip with all four jaws, but if you are attempting to hold hard materials such as metals, the likelihood is that the grip you get with this chuck will not be secure enough for metal turning applications. In this case, it is wiser to use the 4-jaw independent or the 3-jaw self-centering chucks. As with the 3-jaw chuck, a set of four replacement soft jaws is available (#1061).

Figure 12 - 4-jaw Self Centering Chuck and Jaw Set

The chuck adapter

Figure 13 - Chuck adapter

The chuck adapter (#1221) allows a standard Taig chuck to be attached to the milling slide (see *Milling slide and milling vice*), for example to allow a secondary drilling operation to be performed on a part that had been machined on the lathe. Any of the chucks, faceplates, arbors, etc., that carry an internal 3/4"-16 mounting thread can be screwed onto this adapter, and it has the standard Taig internal taper, so can be used with the standard Taig collets.

The face plate

The face plate (#1035) is a simple but versatile work-holding device. The Taig face plate is approximately 3.25" in diameter, threaded to fit the spindle nose, and with two T-slots cut across a diameter; see Figure 14.

Workpieces are attached to the face plate by means of simple strip clamps and angle-brackets (#1036); the existing T-slots in the face plate can be used, or if necessary, holes can be drilled and tapped in the plate to take clamping screws where the T-slots are too limiting. As with many of the Taig accessories, the face plate is made from free-machining steel and can easily be modified in this way. Some people choose to drill and tap a grid pattern of holes on the face plate to minimize set-up time later.

Figure 14 - The face plate and brackets

If turning between centres is to be done, the face plate can be pressed into service as a catch plate to drive a "driving dog", as described further in *Turning between centres*.

Care must be taken when using the face plate, to ensure that the resultant set-up is reasonably well balanced. If necessary, counterweights can be clamped to the face plate to reduce vibration that could otherwise damage the spindle bearings or lead to a poor surface finish.

Tailstock drill chucks

The optional drilling tailstock (#1150) has a ram that is threaded 3/8"-24 TPI to accept Jacobs-type drill chucks. By fitting a drill chuck to the ram, the lathe is converted into a sensitive drill-press that will allow holes to be drilled on-axis in a work piece held in the lathe chuck.

Taig offer four drill chucks, with capacities from 1/4" to 1/2", and in industrial or commercial quality (#1090 through #1093); other drill chucks with a female 3/8"-24 thread should also fit the tailstock ram. The range of chucks can be seen in Figure 15.

The ability to drill from the tailstock is the single most persuasive reason for buying a tailstock. As will be seen later, the tailstock has other useful attributes; however, I would guess that 95% of my use of the tailstock to date has been with a drill chuck attached.

Figure 15 - Tailstock drill chucks

Drill chuck arbor

Figure 16 - Drill chuck arbor

It is also possible to use a Jacobs chuck in the headstock, by using the drill chuck arbor, shown in Figure 16. This is useful for drilling holes in items clamped to the cross slide of the lathe or held in the vertical slide, as also shown in the centre photo. The end of the arbor is threaded 3/8"-24 to take a standard Jacobs drill chuck with the same female thread. The left hand photo (#1140[3]) shows the arbor for the

[3] This is an updated, and rather more robust, arbor than Taig used to offer. As the old design relied on the spindle's internal taper to keep it from slipping, it was not suitable for heavy use. The new design is much simpler, easier to fit, and capable of heavier work.

standard Taig spindle nose; the right hand photo (#1140ER) shows a 3/8" shank arbor designed for use with the ER16 spindle.

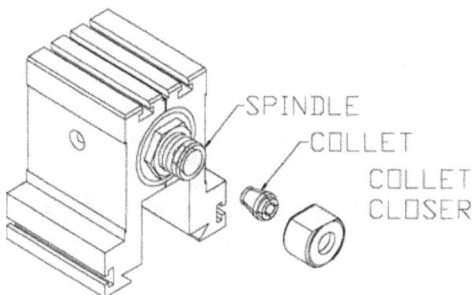

Figure 17 - Fitting a Taig collet

Where standard round stock sizes are being used, and a higher degree of concentricity is needed than can be achieved using the 3-jaw chuck, the Taig collet set (#1040) can offer a useful alternative. The collet set is made to a proprietary Taig design, and is very cost effective; the set consists of a collet closer that screws onto the spindle nose, seven sizes of collets from 1/8" through 5/16" in increments of 1/32", and one blank collet that can be machined for special purposes. Figure 17 shows how the collets are inserted in the spindle taper; Figure 18 shows the full collet set, and a collet in use to machine a length of round stock.

Figure 18 - Taig collet set

The gripping range of each collet is not very great; they are really only useful for gripping stock that is nominally the same diameter as the collet's bore.

As well as holding stock to be machined, collets are the recommended way of holding milling cutters in the lathe headstock.

The blank collet (available separately in sets of 4, #1043) has a small centre mark machined in its outer end to make drilling the collet easier from the tailstock; it shouldn't be necessary to use a centre drill to start the hole before using the final drill size. Bear in mind when drilling the blank collet that the spindle bore is a shade under 3/8", so this places an upper limit on the size of hole that can be drilled in a blank collet without damaging the spindle.

WW collet headstock

If even greater accuracy is desired, it is possible to purchase an alternative headstock for the lathe, bored and tapered to use "WW" watchmaker's collets (#1020). These collets are very close in size to the 8mm collets used in European watchmakers' lathes[4].

ER16 collet headstock

It is possible to use the Taig Mill headstock (#2100H) in the lathe; this headstock has a spindle that is designed to accept industry standard ER16 collets, and Taig offer a set of these collets (#1040ER – see Figure 19) to suit. An advantage of ER16 collets is that they are available in a larger range of sizes than the standard Taig collets, up to 10 mm diameter, and the spindle will accept stock of this diameter. ER collets also have a wider gripping range than the Taig collets; they can grip stock that is up to 1mm smaller than their nominal diameter.

Figure 19 – ER16 collets

---

[4] I suspect that WW collets have a shank that is nominally 5/16" diameter, which is 7.9375mm, as opposed to the true 8mm diameter shank of the European version. The thread used for the drawbar also differs between the two collet standards, one being metric, the other Imperial.

The ER collets have an annular groove at the top, as can be seen in the picture; this groove engages with a spring clip inside the collet closer, so the collet must be fitted to the closer before the closer and collet are fitted to the spindle. The spring acts as a collet extractor that frees the collet from the spindle taper as the collet closer is unscrewed.

Versions of the 3-jaw self-centering chuck and the 4-jaw chuck are available that will fit the ER16 spindle (#1050ER and #1030ER).

5C collet headstock

Figure 20 - 5C collet headstock

A recent addition to the work holding options for the Taig lathe is a 5C collet headstock (#400-00, Figure 20), developed originally for the Taig Turn range of CNC and manual lathes, but which will also fit the Taig lathe bed with the spindle on the same centre height as the other headstocks. The 5C collet headstock accepts industry standard 5C collets; these can be obtained in a wide range of sizes, from 3mm to 26mm in Metric and 1/16" to 1" in Imperial. This is something of a game changer for the Taig lathe, as it offers the ability to grip standard stock sizes up to 26mm/1" in diameter, with a through bore of that size, greatly increasing the versatility of the lathe. Amongst the possibilities here are that a 5C collet could be used to hold an ER-16 plain shank collet chuck[5], or a standard Taig spindle, and in turn, the other chuck and collet systems could be supported within the same machine.

---

[5] In fact, the through bore could accommodate an ER-25 or ER-32 plain shank collet chuck, and other plain shank tooling, so the possibilities here are enormous.

ER16 collet adapter

It is possible to use ER16 collets with a standard Taig lathe headstock by means of the ER16 adapter (#1045, see Figure 21), as an alternative to using the Er16 headstock (see *ER16 collet headstock* on page 33). The adapter comes in two parts; the collet holder, on the left of the first photo, with an internal taper to fit an ER16 collet, is screwed onto the standard Taig spindle, and the collet nut, which tightens the collet into the collet holder. As with the ER16 collet headstock, the collet must be fitted into the collet nut, with its groove properly engaging the internal clip, for this to work properly and for the collet to release correctly when the collet nut is unscrewed.

Figure 21 - ER16 adapter

Blank and special purpose arbors

It is sometimes desirable to make up special work-holding or tool-holding devices for use in particular machining operations, where use of chucks or a face plate doesn't produce a convenient method of holding the work piece or tool concerned. To help in this process, Taig supply a variety of "blank arbors" that can be machined and adapted to suit. These arbors are made from 1" diameter free-machining steel and are bored and threaded so that they screw onto the spindle nose. There are a number of variants:

THE TAIG/PEATOL LATHE

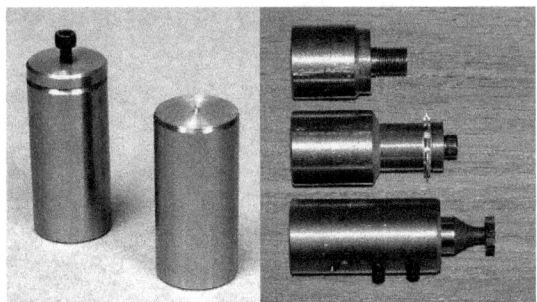

Figure 22 - Blank arbors

— A plain blank arbor (#1132), approximately 2" long, with the free end completely un-machined. This can be seen at the left of Figure 22. There is a variant (#1132ER) that screws onto the ER16 spindle nose instead of the standard Taig spindle.

— A capped blank arbor (#1130), as per the plain arbor, but drilled and tapped axially, and supplied with a capping washer and hex head screw. This can be seen to the right of the plain blank arbor in Figure 22.

— A slitting saw arbor (#1181); this is simply a variant of the capped arbor (#1130), machined at the end to hold a slitting saw.

— A grinding wheel arbor (#1120), again 2" long overall, where the last 7/8" has been machined down to a 3/8" diameter spindle, and about the last 1/2" of that has been threaded. This is supplied with a 5/32" thick backing washer, 1 3/8" diameter, and a 1 3/8" diameter round "nut" tapped to match the threaded portion of the arbor.

The grinding wheel arbor is intended, as its name implies, to be used with the grinding wheel set (#1122, see *Grinding wheel set* on page 60); however, it can also be pressed into service or modified for other purposes along with the other blank arbors.

The right-hand half of Figure 22 shows three blank arbors, machined by the Author for various purposes. From top to bottom:

— A Jacobs chuck adaptor that will screw onto the spindle nose. The arbor was first turned down to 3/8" diameter, then threaded to produce a 24 TPI thread. I made this adaptor primarily for use in the Taig mill, where, because of the dimensions of the machine, using the normal drill chuck arbor is a little awkward. This was machined from a plain blank arbor (#1132).

— A holder for multi-tooth gear cutters, with one of the cutters fitted. This was machined from a capped blank arbor (#1130). The machining consisted simply of turning down the end of the arbor to the internal diameter of the cutter, for a length slightly less than the cutter thickness; the end cap then nips the cutter and holds it firm.

— A milling cutter holder, to hold 1/2" diameter shank milling cutters. It is shown fitted with a small T-slot cutter. This was machined from a plain blank arbor (#1132) by drilling/boring it axially to just under 1/2" diameter, finishing with a 1/2" reamer, and then cross-drilling and tapping for a pair of UNF 10-32 set screws.

— A further example is shown in Figure 23. On the left is the dividing head described in "Dividing and graduating" on page 93, made from a spare Taig lathe spindle, bolted to the bed of a Taig CNC mill. The dividing head has a modified capped blank arbor (#1130) screwed to its nose; this holds a brass blank for a clock gear wheel. The wheel arbor was made by turning down a blank arbor to the diameter of the wheel's central hole (3/16" diameter), and then threading the end of the arbor to take suitable nuts. The wheel is clamped in place with washers either side to prevent cutting forces from distorting the gear wheel. The fly cutter used to cut the gear teeth is held on the nose of the Taig mill spindle (identical to the lathe spindle) using the same milling cutter holder shown at the bottom of Figure 22.

Figure 23 - Another use for a blank arbor

The range of possible uses for blank arbors is very large, and as they are cheap to buy, they make very useful additions to the Taig user's tool kit.

Tailstock support

When using face plates and chucks to hold work pieces in the lathe, it is very important to ensure that the work is held securely, and that the cutting forces operating on the work when turning is in progress will not dislodge or deflect the work piece. It goes without saying that it is not a smart move to be near to a chunk of spinning metal that is likely to come free from the chuck!

Sometimes, the nature of the turning operation is such that it is only possible to grip a short length of the work piece in the chuck relative to its diameter and length; for example, when one end of the work is to be machined and the other end only has a narrow shoulder that can be gripped, or where the diameter is large relative to the length that can be gripped. In these cases, tailstock support may be needed to ensure that the work piece will not come free as machining progresses.

If the work piece is very long, the cutting forces operating on the unsupported end may be enough to deflect the work piece, leading to the cut depth changing over the length of the work. Again, tailstock support or the use of a steady may be necessary in this case.

The use of the tailstock for these purposes is dealt with in more detail in "Turning between centres" on page 70 and "The lever operated drilling tailstock" on page 54.

Chapter Four
Cutting tools

The subject of cutting tools for the lathe is vast, and a detailed treatment of them all is outside the scope of this book. Hence, I will concentrate on the more commonly used, and useful, types.

Tool posts

Taig tool posts come in two varieties:

The normal tool post (#1170), anodized a golden colour, holds 1/4" square lathe tools in the "conventional" working position, where the tool is at the front of the lathe, between the operator and the work piece, and is moved away from the operator to increase the depth of cut, as shown in Figure 24.

The cutting tool is mounted in the tool post so that its cutting edge is at the top, and is on the centre line of the lathe; i.e., the body of the tool is mostly below the centre of the work. If necessary, shims are used under the tool to raise the cutting edge to exactly centre height. Adjusting the tool height so that it is on centre can be a achieved by a variety of means; one of the simplest is to take a facing cut across the end of a piece of round stock, and measure the diameter of the "pip" left in the centre of the cut. The diameter of the pip is twice the amount by which the tool is below centre; shims of half that diameter placed under the tool will bring it to the proper working height, and this can be checked by taking a second facing cut that should leave no pip.

Figure 24 - The standard tool post

A simpler, but less accurate, method of setting tool height is to use the point on the end of the tailstock ram as a height gauge, bringing it up to the tool and comparing it by eye with the position of the cutting edge. If necessary, the tool post can be loosened and rotated to make this possible[6].

A useful quick and dirty "trick" that can help with adjusting the height of a cutting tool is to lightly trap a strip of flat metal, such as a small engineer's rule or a strip of thin metal from the scrap box, between the tip of the cutting tool and a piece of bar stock held in the self-centering chuck. If the top of the strip leans towards you, the point of the tool is too low; if it leans away from you, the point is too high; if it stands vertically, it is just right.

Figure 25 - Front and rear tool posts

Working with a tool that is not set correctly at centre height will lead to a poor cutting action, resulting in a poor surface finish, and the danger that the tool may "dig in" to the work piece, potentially damaging the tool, the lathe, and the work.

Taig tool posts are cheap accessories; I find it useful to have as many tool posts available as lathe tools that I use regularly. That way, I can leave the tool already shimmed to height in its own tool post. Alternatively, it is worth keeping the right set of shims with each tool, perhaps by holding them together with elastic bands, or by keeping each tool with its shims in an old 35mm film cannister[7].

---

[6] Clearly, this only works if the tailstock is accurately on centre height, which it may not be. Checking this is discussed in *Checking alignment* on page 73.

The Taig rear tool post (#1171) is black anodized, and is designed to be used behind the work, with the tool mounted upside-down. The rear tool post can be seen contrasted with the front tool post in Figure 25; the rear tool post is somewhat taller, and the slot for the tool is higher, to allow the cutting edge to be at centre height with the tool mounted upside down. Again, shims are used under the tool to adjust its height correctly.

The most common use for the rear tool post is to hold parting off tools. Part of the reason for this is that parting off tools, because they cut over a relatively wide area, exert large cutting forces that would tend to cause the tool to dig in to the work if the tool is used in the front tool post. In the rear tool post, the cutting force will tend to deflect the tool away from the work, thereby reducing the depth of cut and avoiding the tendency for the tool to dig in.

The rear tool post can also be used with other lathe tools. Either way, it can be convenient to mount two tools in the lathe at the same time, so that two different cutting operations can be performed without changing tools. This is particularly convenient if batches of identical parts are being made.

A variant of the rear tool post, the T-bar Cut Off Tool (#1173), provides a complete parting off tool with its own T-section blade.

Figure 26 - T-bar cutoff tool

When using any parting off tool, be sure to use lots of cutting oil to lubricate the cut and ensure that the cutting edge is properly adjusted and sharpened. Parting off is a skill that must be worked at[8]. The most common problem is that the parting tool "chatters" in the cut, leading to a poor-quality cut; if this is happening, and the tool is at centre height and sharp, try reducing the lathe speed and increasing the rate of cut.

"Quick change", or QC, tool posts can be obtained in sizes that are suitable for the Taig lathe. However, as removal and re-fitting a Taig tool post is a matter of loosening a single T-bolt, I have never really seen the point of buying into a more expensive tool holding system. The only real advantages that QC tool posts seem to have is that the tool height can be adjusted without shims, by means of a lockable adjusting screw, and because the removable tool holders slot into a fixed tool post, a

---

[7] "What's a 35mm film cannister, Grandpa...?"

[8] Of course, "Parting is such sweet sorrow", according to The Bard...

tool can be removed and re-fitted knowing that it will be returned to exactly the same position it was in before. However, I have yet to find the lack of either feature at all limiting in my use of the lathe.

HSS tool bits

Taig supply a set of lathe tool bits (#1095), made from high speed steel (HSS), as seen in Figure 27. These tools are available individually (#1097 A through F). Unground 1/4" square blank tool bits are also available (#1096); one is shown at the bottom of the photograph.

Figure 27 - HSS lathe tool set

The set of six tools comprises (from left to right of the photo):
— A boring bar, used for boring into a work piece. Boring bars need a starter hole in order to work; the diameter of the starter hole will depend on the size and geometry of the tool. The Taig boring bar in its unmodified form needs starter hole that is a little over 1/2" in diameter.
— A cutoff tool, intended for use in the rear tool post.
— A right hand knife tool, used to cut from right to left up to a shoulder, and can also be used for facing cuts on the right hand side of the work piece.
— A left hand knife tool, used to cut from left to right up to a shoulder, and for facing cuts on the left of the work piece.
— A 45-degree chamfer cutting tool. This allows a chamfer to be cut very simply on the end of a part, without having to resort to the use of the compound slide (see *Compound slide* on page 57).
— A round nosed tool, useful for "roughing out" parallel cuts, and also for cuts where rounded corners and contours are required.

HSS tool bits are suitable for work on non-hardened materials, such as aluminium, mild steel, brass, and so on. Sharpening these tool bits is simple, using conventional white or grey grit abrasive wheels on a bench grinder. Once mastered, re-sharpening can be achieved quite quickly. Sharp tools are essential for easy machining and good surface quality, so this is a skill worth learning. *Sharpening lathe tools* on page 45 describes some simple techniques for this, using the right hand knife tool as an example.

An unground ¼" square tool bit is available for the user to make special purpose cutting tools or form tools (#1096).

Carbide tools

Carbide tipped tools can be obtained in sizes that will fit the Taig tool posts (1/4" or 6mm shank), from the usual machinery and workshop suppliers. These tools have a small piece of tungsten carbide brazed onto the end of a steel shank; the carbide tip is sharpened to form the cutting tool, with the steel shank serving solely as a support.

The advantage of carbide tipped tools is that they will tackle harder materials; for example, cast iron castings, where the outer skin can be very hard and difficult to penetrate with HSS, or stainless steel where it is easy to work harden the machined surface if the cutting rate is too low. Carbide tools can also be used at significantly higher spindle speeds, up to 3 times the speed used for HSS.

Replaceable tip tools

Replaceable tip tools, sometimes known as "indexable" tools, are designed to carry carbide inserts that have pre-sharpened edges. These inserts come in a wide variety of shapes for different purposes, but the ones most commonly used in lathe tools are either triangular or diamond-shaped, and provide 2, 3, 4, 6 or 8 cutting edges, depending upon how the insert and the holders have been designed.

The great advantage of these tools is that when the edge you are using becomes worn, you can turn the insert to reveal a new cutting edge, or replace the insert, and know that the cutting edge is still on centre height. The main disadvantage is higher cost, as once the edges have been used up, the tip is discarded as it is generally not practical to re-sharpen them. Figure 28 shows a selection of these tools.

From the top down, these are:
— A small boring bar with a triangular tip, allowing holes as small as 1/4" diameter to be bored.

- A slightly larger boring bar.
- A right-hand roughing and facing tool. This uses the same inserts as the second boring bar but uses the obtuse-angled cutting edge of the insert. Hence, inserts that have had the acute-angled cutting edges worn out in the boring bar (or the right-hand knife tool below) can be fitted in this tool to use their remaining cutting edges. These edges are rather stronger than the acute-angled edges and are therefore suitable for heavier roughing cuts.
- A "button" tool, using a carbide insert with a circular cutting edge. This is useful for turning curved contours or scalloped chamfers; however, as it cuts over a considerable width of the insert, it must be used carefully, and with good lubrication, to avoid chatter.

Figure 28 - A selection of "indexable" tools

- A right hand knife tool, using the same diamond-shaped insert as the facing tool and boring bar. This can be used to turn up to a shoulder.
- A cutoff tool, designed for use in the rear tool post.

The inserts are supplied in a variety of tip radii; the larger the radius, the higher the cutting force needed to use the tip effectively; conversely, the smaller the radius, the slower the feed needs to be in order to generate an acceptably smooth surface. This latter point is easily understood when you realize that the lathe tool actually cuts a spiral path along the length of the work piece. Effectively, what is being formed is a fine screw-thread whose pitch is determined by the rate of movement of the saddle. To get the best possible surface finish, the tool should move a distance of twice the tip radius or less for each full revolution of the spindle. So, for example, if machining 1/2" diameter brass at 4500 RPM with a 0.2mm radius tool, a saddle traverse rate of about 2 yards/minute (about 1.2"/sec) would be the maximum for a good finish, and preferably much less.

A tip radius of 0.1mm is great for producing very fine surface finishes (it can produce almost mirror finishes on aluminium and brass); 0.2mm seems to be a good general purpose tip radius on a lathe this size, and anything much larger than 0.4mm is OK for roughing cuts in soft materials, but is getting to be too large for use with the harder materials.

Indexable tool holders with 1/4" shanks (or sometimes, 6mm shanks, which is very slightly less than 1/4") used to be difficult to obtain, but with the increasing popularity of small lathes, are now more readily available. See "Tool suppliers" on page 171 for some UK-based suppliers of these tools.

Sharpening lathe tools

Tools made from HSS can be sharpened using a conventional bench grinder. The simplest lathe tool to sharpen is the right-hand knife tool. A little study of one of Taig's pre-sharpened tools will reveal the general principles:

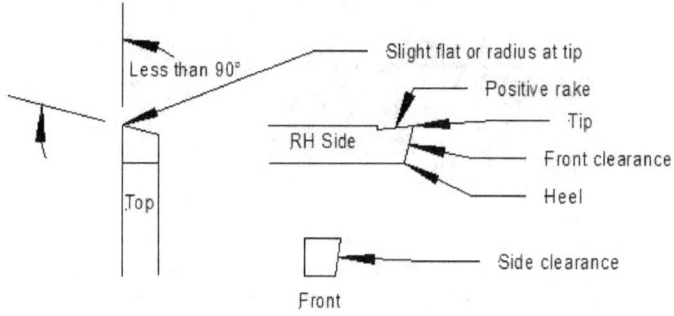

Figure 29 - Grinding a right-hand knife tool - 1

— For the tool to be able to cut up to a square shoulder, it is necessary for the angle between the two edges of the tool (as viewed from above) to be less than 90 degrees.
— A certain amount of "positive rake" helps the cutting edge to cut into the material being turned. In a positive rake cutting tool, the top surface of the tool is angled up towards the work piece. However, with some materials, for example, brass, zero or negative rake is preferable as positive rake tools have a tendency to "grab" the surface and cause the tool to dig in.[9]

---

[9] "Grab" can also be an issue with twist drills; some machinists that regularly drill brass will keep a set of twist drills with the cutting edges deliberately stoned off

— Side and front clearance is needed to ensure that the heel of the tool doesn't rub against the work piece.

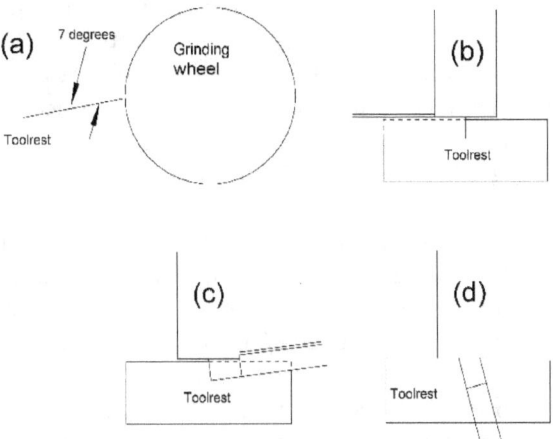

Figure 30 - Grinding a right-hand knife tool - 2

— Although at first glance, the point of the tool appears to be just that (a point), a well-ground lathe tool will have a slight radius or a small flat at the tip. This improves the surface finish that the tool will produce.

These requirements are illustrated, in simplified form, in Figure 29.

A close approximation to this tool form can be achieved relatively simply, as follows (with reference to Figure 30 for illustrations):

— Adjust the tool rest on your bench grinder so that it makes a 7 degree angle to a radius centered at the grinding wheel spindle as in Figure 30 (a), and dress the wheel to ensure that the wheel presents a fresh and flat cutting face.
— Grind the side clearance by holding the tool bit as in Figure 30 (b), and moving it from side to side against the wheel until a sharp edge is formed along the top of the tool bit.
— Grind the top rake by holding the tool bit as in Figure 30 (c), continuing the grinding until the edge is sharp.
— Grind the front clearance by holding the tool bit as in Figure 30 (d), with the top surface of the tool uppermost, until the end is ground all the way to the tool tip.

---

to give zero cutting rake for drilling such materials.

— Form a slight flat or radius from the tip of the tool to the heel, either by careful use of the grinding wheel or by using a slip stone.

This will give a result similar to the cutting tool shown in Figure 29, with the exception of the top rake, which is slightly tilted relative to that shown in the drawing. While this produces a less-than-ideal tool form, it is a good compromise, as the simplicity of the setup needed outweighs the slight reduction in performance relative to the ideal form.

It is worth having a pot of water handy to cool the tool bit while grinding, not so much because there is any danger of damaging the tool through overheating (although this is a possibility if the tool becomes excessively hot, so it is not advisable to allow the tool to become so hot that visible colour changes occur), but more to avoid burnt fingers when handling the tool. Also bear in mind that the edges produced by grinding are likely to be very sharp and quite capable of cutting flesh, so the tool should in any case be handled with care.

If the above instructions are followed, then the result is three ground surfaces that are flat other than the slight curvature imparted by the grinding wheel. If a particularly fine finish is required, the three ground faces can be honed using a slip stone in order to generate a very fine cutting edge; however, for most purposes, the edge produced by the grinding wheel is good enough for immediate use.

If the tool is to be used for brass, omit the grinding of the top rake, and simply stone off the top of the tool with a fine slip stone to produce a flat surface.

Grinding a left hand knife tool is performed in a similar manner, but with the opposite "handedness". The 45 degree chamfer tool is simply a right hand knife tool with the point ground off at an angle of 45 degrees to the shaft of the tool.

Boring bars

Boring bars are, as their name implies, used to bore holes in work pieces, concentric with the centre of the lathe spindle. As mentioned earlier, these tools are usually not capable of "plunge cutting" and require a pilot hole that gives appropriate side clearance to the cutter.

THE TAIG/PEATOL LATHE

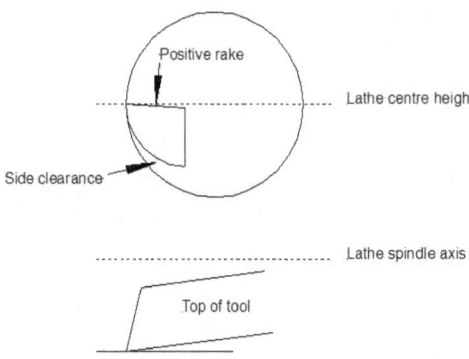

Figure 31 - Taig boring bar clearances

Because boring bars are generally used with a significant "overhang" (the distance from the cutting edge to the tool post), it is important to keep the tool as rigid as possible, consistent with the requirement to bore holes of a particular diameter. Hence, while a boring bar designed for boring small diameter holes is capable of boring larger diameters, it will inevitably be less rigid than a larger diameter tool and will therefore only be capable of taking lighter cuts. It is therefore useful to have a range of sizes of boring bar to cover the range of diameters that you are likely to work with.

The Taig boring bar provided in the 6-piece tool set has a cutting edge formed from a quadrant of a circle approximately 1/2" in diameter (see Figure 31), with a little positive rake applied to the upper surface, and the leading edge of the tool ground for clearance, in a similar manner to the right hand knife tool. As the quadrant section of this tool is parallel, it should be used at a slight angle to the axis of the lathe in order to prevent the tool rubbing in the bore as the cut proceeds. Hence, in its supplied form, this tool needs a starter hole that is at least 1/2" in diameter and preferably a little more, although it is possible to grind the tool to provide the clearance necessary for use with smaller diameter holes if need be. As with the other turning tools, the cutting edge is set at centre height by the use of shims.

Making a boring bar holder

Many commercial boring bars have a round section shank and are difficult to hold securely in a normal Taig tool-holder. It can be convenient to make up a simple boring bar holder that can carry a range of shank diameters, as shown in Figure 32 .

The holder is made from a piece of aluminium bar, 1.5" X 1" in cross-section and 1.75" long. The two ends of the bar are cleaned up in the lathe by holding the piece in the 4-jaw chuck and making a facing cut across each end. The bar is then drilled for the 7/32" hole and is attached to the cross slide using a UNF 10-32 T-bolt and square nut, in a similar manner to the standard tool post, with the wide face at right angles to the axis of the lathe. If necessary (i.e., if your T-bolt isn't long enough!), the 7/32" hole can be recessed at the top to take the diameter of the T-bolt head.

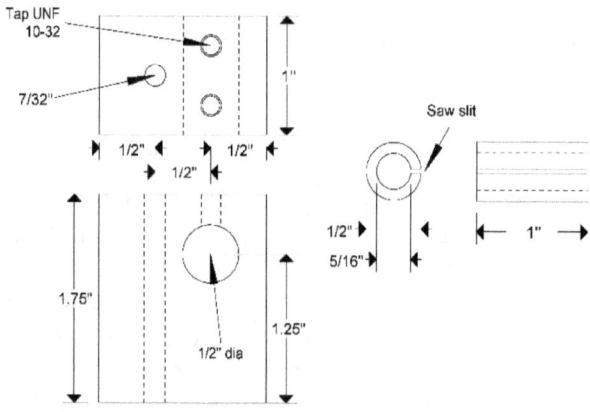

Figure 32 - Boring bar holder - drawings

A 1/2" drill chuck is fitted to the lathe spindle by means of the drill chuck arbor, and a pilot hole drilled for the 1/2" diameter hole, by using a centre drill. The 1/2" hole is probably best drilled in stages, using progressively larger drills. The two holes for the set screws that will hold the boring bar in place are drilled and tapped UNF 10-32; a 4.1mm drill is a suitable size for tapping these holes.

The collet shown on the right of Figure 32 can now be made up by cleaning up the ends of a length of 1/2" diameter brass rod in the 3-jaw chuck to give a 1" finished length, and then drilling the brass axially from the tailstock to the same diameter as the boring bar shank. If more than one shank diameter is to be used, then a collet can be made for each diameter. The collet is slit along its length with a hacksaw, to allow the pressure from the set screws to make the collet grip the boring bar.

Figure 33 - Boring bar holder - completed

Figure 33 shows the finished tool holder, fitted with a commercial "indexable" boring bar that uses disposable carbide inserts. The tool is rotated in the collet to bring the cutting edge exactly to centre height. Boring bars of this type can be obtained that will start from a pilot hole as small as 1/4" in diameter.

Figure 34 - Modified tool post

For small diameter boring bars, a simpler approach is to modify a standard Taig tool post to take a boring bar, as illustrated in Figure 34. Here, the indexable boring bar shown earlier at the top of Figure 28 has been fitted to the tool post by simply drilling a hole through the tool post exactly at centre height (using a Jacobs chuck to hold the drill in the lathe spindle), and using a drill of the same diameter as the boring bar shank. As before, two set screws are fitted to clamp the bar. Commercial indexable boring bars are often manufactured with a flat on the top surface of the shank; set screws bearing down on this flat will hold the tool such that its cutting edge is at centre height if the tool post has been drilled at centre height with the right diameter of hole. However, when using this approach, bear in mind that there is not much "meat" to drill into between the central T-bolt hole and the edge of the tool post, so it is only suitable for boring bars with shanks of around 1/4" in diameter or less.

Radius/ball turning tool

Figure 35 – Radius/ball turning tool

The radius tuner (#1210), often known as a ball turning tool, allows a lathe tool to be moved in an arc, thereby cutting a radius on the workpiece. Figure 35 shows this device; it is clamped directly to the cross slide by a T-bolt that passes through the centre of a large ball bearing (actually, the same as the ball bearings that is used in the headstock). The tool holder rotates with the outer component of the ball race. The distance between the tip of the cutter and the centre of the T-bolt determines the radius that the tool will cut; this can be adjusted by means of the two set screws that hold the cutter in place.

As with all lathe tools, the tip of the cutter should be kept at centre height. The cutter is a symmetrical round-nosed tool; fitting an asymmetric cutter (a left- or right hand knife tool for example) won't work with this device, as the cutting tip will not be aligned with the centre of the bearing.

Slitting saw

The left-hand photo in Figure 36 shows the slitting saw arbor (#1110) and the slitting saw (#1111). The arbor screws directly onto the spindle nose; it has a removable cap that is held by a central cap-head screw. The slitting saw is placed over the spigot on the end of the arbor, and the cap screws on to hold it in place. The right-hand photo shows a variant of the slitting saw arbor (#1110ER) intended for use with ER16 collets – it has a 3/8" diameter shank that is held in a 3/8" diameter ER16 collet.

The slitting saw can be used to make accurate cuts in parts clamped to the cross slide. However, as the saw is 2.5" in diameter, the parts have to be clamped in such a way as to allow them to overhang the edge of the cross slide.

Figure 36 - Slitting saw

When using a slitting saw, bear in mind that it is a large, multi-tooth cutter, and should be used at much slower spindle speeds than would be appropriate for turning with a single point cutting tool. It is also worth using cutting fluid or oil to keep the saw cool and reduce wear.

Considerable care should be used with these saws, as there are very capable of cutting flesh and bone, and no protection is offered to the exposed blade when they are in use!

Using cutting fluid

It is advisable to use a cutting fluid or coolant on most materials, as this will improve the tool life and the surface finish. There are some minor exceptions however; brass, particularly the leaded "engraving" brass that is used in clockmaking, cuts very freely without the use of a cutting fluid, and this does not appear to reduce tool bit life.

There are many proprietary cutting fluids available for use with the lathe. Some of these are water-based and are best avoided unless the lathe is carefully dried off and re-oiled after use; neat cutting oils are more appropriate for use with this lathe.

The cutting fluid can be applied using an oil-can, or periodically brushed directly onto the surface of the work with a small paintbrush.

For aluminium, WD-40 or paraffin (kerosene) works very well; WD-40 is particularly convenient as it can be sprayed directly onto the work. The chips produced when cutting aluminium and some other soft metals have a tendency to adhere to the top edge of the cutting tool; they literally weld themselves to the tool under the influence of the cutting forces and the local heat generated by the cutting action. A cutting fluid will help prevent this buildup, which progressively reduces the tool's cutting performance.

For the more ambitious user, it would of course be possible to set up a pump feed or drip-feed coolant system, and to collect and re-cycle the coolant via a drip tray under the lathe. However, for most purposes, this is an unnecessary complication when working with a lathe of this size.

Chapter Five
Standard accessories

There are many standard accessories available for the lathe, and some that were intended for use with the Taig Mill that can usefully be adapted for use in the lathe.

The lever operated drilling tailstock

The Taig lever operated drilling tailstock (#1150) is shown in Figure 37, where it is being used to drill a hole in a workpiece; however, drilling is not the only use for it.

Figure 37 - The tailstock

Other major uses for the tailstock are to support work that is held in a chuck or "between centres", and to cut tapers (see *Turning between centres* on page 70 and *Taper turning* on page 72).

As can be seen in the photo, there is a socket head screw at the front of the tailstock that allows it to grip the lathe bed at the desired position, a second screw on the top that clamps the tailstock ram when necessary, and a third screw at the back (partly obscured by the operating lever) that clamps the top half of the tailstock to its own transverse dovetail slide[10]. This last screw allows the tailstock ram to be offset from the lathe axis when taper turning, or to be locked on-axis for parallel turning with tailstock support and for drilling operations.

Needle bearing centre

The standard tailstock ram can be removed and replaced by a needle bearing centre (#1151). The advantage of this centre over the standard tailstock ram is that the centre rotates with the workpiece in its own needle roller thrust bearings. Figure 38 shows the needle bearing centre, and also shows it in use to provide tailstock support for a long workpiece.

Figure 38 - The needle bearing centre

Removing the standard "fixed centre" tailstock ram is achieved by removing the split pin peg that connects the lever arm to the ram, loosening the clamping bolt on top of the tailstock and sliding the ram out. The needle bearing centre is then slid into the tailstock in its place and clamped in the required position.

---

[10]Earlier versions of the tailstock had a different slide locking arrangement, using a second locking screw on top of the tailstock. This screw had to be loosened and then tapped down to release the slide.

Milling slide and milling vice

For those that do not have access to a milling machine, it is possible to undertake light milling operations in the lathe by fitting the vertical milling slide to the lathe cross slide.

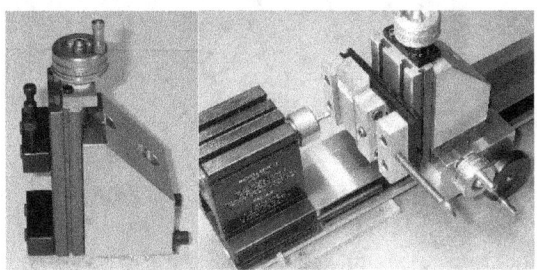

Figure 39 - The milling slide and vice

The milling slide (#1220) is shown in Figure 39. A simple clamping arrangement is provided by two adjustable blocks and clamping screws, as seen in the left-hand photo; these can operate as a crude milling vice. However, a lightweight milling vice (#1225), designed to fit the T-slots in the milling slide, is also available, as seen in the left-hand photo of Figure 40. The vice seen in the right-hand photo (#2225) is essentially the same vice but with a hand-crank fitted in place of the thumbscrew. The right-hand photo of Figure 39 shows the thumbscrew version of the vice in use with the milling slide.

Figure 40 - The milling vices

Milling cutters

A range of double-ended end mills (#1230 A through E) is available; these are 3/16" shank high speed steel cutters and can conveniently be held in the 3/16" collet for milling operations (see Figure 41). They come in a range of cutter diameters from 1/16" to 3/16", in increments of 1/32".

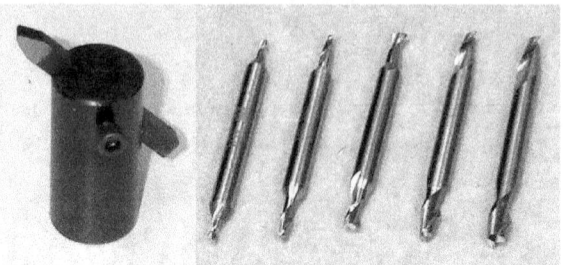

Figure 41 - Milling cutters

Of course, end mills sourced from the usual tool suppliers can also be used, bearing in mind that the maximum collet size using the Taig collets is 5/16", the ER16 collets can hold up to 10mm. If it is necessary to use larger cutters, a cutter holder can be made by boring a blank arbor to suit the cutter shank size and using a set screw to clamp the cutter securely.

A fly cutter (#1224) is shown in the left-hand photo; this takes standard 1/4" shank tool bits. The fly cutter is another example of what can be done with the Taig blank arbors. The body of the fly cutter is essentially a blank arbor that has been bored and fitted with a set screw. Commercial "after-market" fly cutters can also be accommodated by boring a blank arbor to suit the shaft diameter of the fly cutter and cross-drilling to fit a set screw.

Although it may be tempting to use the drill chuck arbor and a Jacobs chuck to hold milling cutters, this is definitely not to be recommended, as this setup is not designed to take the side loads that are generated by milling, and damage to the spindle taper may result. Jacobs chucks are also not capable of holding a milling cutter securely.

Compound slide

The compound slide (Figure 42) is the second means of cutting tapers on the Taig lathe. It is fitted to the cross slide by means of a clamp that slots into one of the T-slots of the cross slide, and locates in a circular dovetail in the base of the compound slide. A single socket head screw fixes the slide in position at the desired angle. Fitting the slide can be a little fiddly, as it entails using a long hex wrench, slid into the T-slot from the far side of the lathe in order to adjust its position; hence, replacing the clamping screw with a knurled knob on an extended screw shank is a useful improvement that can be made to the lathe (see *Locking knobs and handles* on page 89).

Figure 42 - The compound slide

The compound slide has its own tool post - a simple clamp that holds the lathe tool onto the top surface of the slide, as the thickness of the slide makes it impossible to use the standard tool holders. As with the standard tool posts, shims are used under the tool to achieve the correct tool height.

Steady rest

The steady rest (#1190) provides a further option for supporting long workpieces, as an alternative to using tailstock support. As can be seen in Figure 43, the steady rest is clamped to the lathe bed and has three adjustable fingers that can be brought into contact with the workpiece. These fingers are made of brass and are rounded at the ends to prevent them from marring the surface of the work; in use, it is necessary to lubricate the point of contact with the work.

Figure 43 - Steady rest

The steady rest can be positioned at either side of the cross slide, depending upon which portion of the workpiece you wish to machine. In either case, the fingers should be set such that they hold the end of the workpiece exactly on centre.

An aid to achieving this is to fit the steady rest just next to the headstock and adjust the fingers to touch the surface of the workpiece. The steady rest can then be moved to its working position without re-adjusting the fingers, although, if the desired working position is as shown in Figure 43, it may be necessary to remove and re-fit the carriage.

Riser blocks

Occasionally, it is desirable to turn parts that are of larger diameter than can be accommodated in the 2.25" centre height of the lathe. The rising blocks (#1250 and #1251) provide an additional inch or so of centre height, by lifting the headstock, tool post and tailstock.

Figure 44 - Riser blocks

Figure 44 shows part #1250, which consists of a headstock riser block and a tool post riser block; the tailstock riser block (#1251) is identical to the headstock riser block. The headstock is simply unclamped from the lathe bed, the riser block clamped to the bed in its place, and the headstock clamped to the riser block. Similarly, the tool post is clamped on top of the tool post riser block. It should be borne in mind when using riser blocks that the stiffness of the lathe is noticeably reduced. Therefore, you will not be able to make as heavy a cut with the lathe when used in this configuration, so more patience and care is needed.

Chuck depth stop

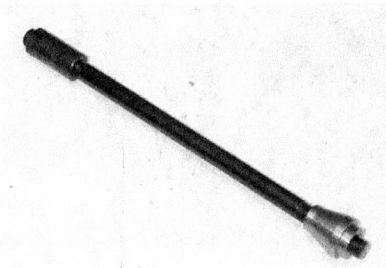

Figure 45 - Chuck depth stop

The chuck depth (#1026) stop is useful when multiple components are being machined to the same length. As can be seen in Figure 45, the depth stop is a blank collet that has been bored and threaded to take a draw bar, and can therefore be fitted to the spindle along with either the 3- or 4-jaw chucks. The draw bar is screwed in and out of the collet to adjust how far the workpiece is able to enter the chuck jaws.

Grinding wheel set

The grinding wheel set (#1122), along with the grinding wheel arbor (#1120) allow the lathe to be used for precision grinding operations (see Figure 46). The set consists of three cylindrical grinding wheels of different sizes that can be screwed onto a grinding wheel arbor.

With care, the grinding wheels can be used to sharpen lathe tools and touch up the cutting edges of end mills and the like.

When using the lathe for grinding, the sliding surfaces and bearings must be adequately protected against abrasive dust from the grinding wheels, and it is advisable to thoroughly clean the lathe after such use. Failure to observe these precautions will result in the lathe slideways and bearings wearing much more rapidly than they would under normal use. Similarly, if you have a bench grinder, it is advisable to keep it well away from the lathe or cover the lathe when it is in use.

Figure 46 – Grinding wheel set

Tailstock die holder

The tailstock die holder (#1152), as its name implies, allows threading dies to be held in the tailstock. The die holder is double ended, accepting 13/16" diameter dies at one end, and 1" dies at the other. The holder is bored and tapped to fit the 3/8" 24 TPI thread on the end of the tailstock ram.

Figure 47 – Tailstock die holder

Spindle wrench

Figure 48 - Spindle wrench

The spindle wrench (#1310) is a simple 1" AF open ended wrench, but the really useful thing is that it is sufficiently thin to fit between the headstock and a chuck attached to the spindle, which is pretty hard to find. It is an essential aid for fitting and removing chucks from the headstock, and in conjunction with a standard 7/8" AF wrench or the ER-16 collet chuck wrench, for tightening the collet closer on the spindle. Perfectly possible to buy a standard wrench and modify it, of course, but for the price, just not worth the effort involved.

Wood-turner's face plate

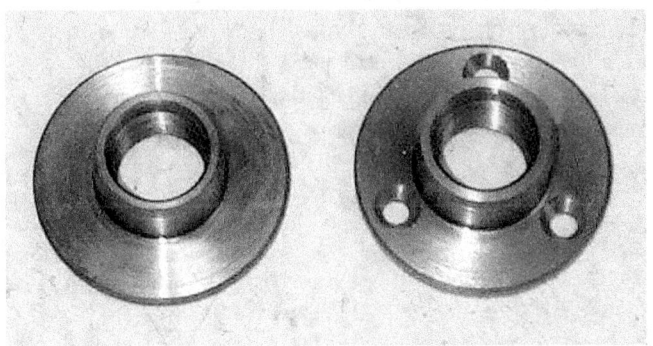

Figure 49 - Wood-turner's face plate

The woodturner's face plate (#1037 and #1037-3H) allows wooden workpieces to be attached to the spindle nose. The face plate comes in two versions, with or without countersunk screw holes. The one shown in the photo allows the workpiece to be attached to the face plate with wood screws; if it is desirable not to mar the wood by screwing into it, the workpiece can be attached to either face plate with a suitable adhesive. A useful approach is to fix a wooden disc to the face plate with screws, then to coat the workpiece and the disc with wood glue and sandwich a sheet of thick drawing paper between the two. The sheet of paper will be strong enough to keep the workpiece attached during turning, but when the turning is finished, the workpiece can be separated from the wooden disc by carefully prising the two apart along the paper join using a sharp knife or chisel.

Adjustable tool rest

Figure 50 - Adjustable tool rest

The adjustable tool rest (#1038) was primarily intended for woodturning applications; however, it can also be used for "freehand" metal turning, for example, using a graver to turn pivots on clock arbors.

The base of the rest clamps directly onto the lathe bed; the rest itself can be adjusted to an appropriate height and distance from the work by means of the clamping screws that can be seen in Figure 50

Lathe dog

Figure 51 - Lathe dog

The lathe dog (#1034) is used to drive a workpiece when turning between centres (see page 58). The simplest way to use the lathe dog is to use it in conjunction with a centre held in the 3 or 4 jaw chuck. The lathe dog is attached to the left-most end of the workpiece; the cap head screw at the top left of the picture acts as a drive pin that engages with one of the chuck jaws in order to transmit power to the workpiece. For safety, the drive pin should be wired to the chuck jaw to prevent it moving independently of the chuck.

Chapter Six
Using the lathe

This chapter is intended to serve as a very basic and very brief introduction to using the lathe, and will therefore be of little interest to the more experienced Taig lathe owner. It is not intended to be an in-depth treatment of turning and machining techniques; rather, the intent is to give the reader enough information to get started, once the lathe has been properly assembled as described in "Assembly and mounting" on page 7. Other, more comprehensive, books on workshop practice are available that deal with machining techniques in small lathes; the reader is strongly encouraged to read around the subject some more before taking the plunge.

Care of the lathe

Chapter 2 discusses cleaning and lubrication of the lathe, so there is no need to repeat that material here. However, there are a few additional points that should be borne in mind to help ensure that your lathe has a long and useful life.

— Some parts of the lathe are prone to rusting; particularly the lathe bed, but also the steel components of the chucks and other accessories. With a small lathe like the Taig, it is unusual to use water-based coolants/cutting lubricants, but if you choose to go that route, make sure that the lathe and its accessories are thoroughly wiped down after use, and then re-lubricated, to prevent rusting.

— If you keep your lathe in a room, garage, or out-house that is not heated and is at all prone to damp or condensation, pay close attention to any parts that are prone to rusting. A thin film of oil over these parts will help to prevent damage. Particularly important in these conditions is to remove the chuck, collet closer, etc., from the lathe spindle after use, particularly if it may be several days or weeks till the lathe will next be used, as it is possible for steel components to become rusted onto the spindle, making it difficult (and potentially damaging) to remove them later.

— Avoid damage to the lathe bed; any damage, particularly to the dovetails, can severely impact both the accuracy of the lathe and the rate at which it wears out. In particular, avoid dropping heavy objects on it, or marking it with cutting tools. It is sometimes tempting (and in my case, I often give in to this temptation!) to use a hacksaw to cut off parts while they are still held in the chuck. If you find yourself doing anything like this, take the elementary precaution of protecting the lathe bed against either the saw blade or the part dropping onto it and marring the bed.

— Abrasives of any kind in contact with the moving parts of a lathe will result in faster rates of wear; the harder abrasives (such as silicon carbide, aluminium oxide, diamond...etc.) can and will embed themselves in metal bearing surfaces, particularly where softer metals are used (such as brass gib strips, aluminium components, and so on). If you decide to use abrasive wheels or materials in or near the lathe, make sure that all possible opportunities for dust ingress are protected, and that the lathe is properly cleaned and re-lubricated afterwards. If you have a bench grinder, linisher, or similar machine that uses abrasive materials, then if you must use your lathe in the same room, site it as far away from these machines as possible, and cover the lathe when they are in use.

— Avoid pushing the lathe beyond its reasonable capabilities. Machines often give the operator fair warning when they are being pushed too hard - they get hot, make strange noises, belts slip, and so on. Take heed of such warnings; they are trying to tell you something you need to know.

— Keep the various slideways, handwheels, etc., properly adjusted. This will improve the accuracy of the lathe and make it easier to use.

— The moving parts of the lathe will inevitably wear in time to the point where adjustment is no longer possible; replacements for all parts of the lathe can be obtained from the Taig factory or their distributors.

Fitting and removing chucks

Figure 52 - Protecting the lathe bed

Before fitting or removing a chuck, it is a wise precaution to protect the lathe bed from potential damage if you were to inadvertently drop the chuck. On larger lathes, the operator is usually advised to make or buy a "lathe board" to protect the bed; in its simplest form, this is a piece of wood the width of the lathe bed, with side pieces that clip it onto the lathe bed.

Although there is nothing wrong with making one of these for the Taig, it is possibly overkill, as a piece of corrugated cardboard cut from an old cardboard carton will do the job much more simply, as illustrated in Figure 52.

Make sure that the threads of the spindle and the chuck are both free of swarf and dirt. The chips produced by machining aluminium and other soft metals can become embedded in the threads if they are not cleaned out before the chuck is fitted; as a result, the threads in the chuck get smaller (or the spindle threads larger), making it increasingly difficult to fit and remove the chuck. The other important reason for keeping these threads, and the chuck/spindle "register" (the short plain-turned portion of the spindle at the base of the spindle thread, and the corresponding socket at the back of the chuck) clean is that dirt or swarf can prevent the chuck from seating properly on the register.

Lightly oil the threads. Thread the chuck carefully onto the spindle nose while holding the spindle drive pulley still with the left hand, making sure that the chuck is kept square to the thread to avoid cross-threading. Tighten the last 1/2 turn with a smart flick of the wrist to seat the chuck tightly on its register.

It shouldn't be necessary to tighten the chuck onto the spindle any more than can be achieved by this means; over tightening may make subsequent removal problematic and may result in damage to the headstock or spindle. Conversely, insufficient tightening can result in the chuck coming loose during machining, so there is a balance to be achieved here.

To remove the chuck, use the spindle wrench (see Figure 48) to hold the spindle still, and lightly tap one of the chuck jaws with a wooden mallet to release the threads.

The 3-jaw chuck jaws are tightened and released by means of a short tommy bar. The chuck is not intended to be tightened more than can be achieved by hand using this tommy bar; applying greater force than this will probably damage the scroll, and certainly bend the tommy bar.

Using collets

The Taig collets offer a useful method for holding round stock in the lathe and will generally centre the workpiece much more accurately than with a self-centering chuck. However, the gripping range of each collet is quite small; the intent is that the diameter of the stock should be the same as, or only slightly smaller than, the corresponding collet size. Attempting to hold stock that is too small will result in damage to the collet.

As with fitting a chuck, make sure that the collet closer and spindle threads are clean and lightly oiled before use. Tightening the collet closer by hand is usually enough to grip the workpiece while you adjust how much protrudes beyond the collet; the collet closer should then be tightened using the spindle wrench (see Figure 48) and a 7/8" AF wrench. It is not necessary to use a great deal of torque to tighten the collet closer, especially with the smaller diameters.

The procedure is slightly different for ER16 collets, as the ER16 collet closer incorporates a spring clip that engages with the slot at the end of the collet; this clip acts as a collet extractor when the collet closer is unscrewed. The collet must be inserted into the collet closer so that it properly engages with this clip before the collet and the closer are fitted to the spindle. Failure to do this will result in the collet being off centre and may damage the closer and/or the collet.

Selecting the right cutting speed

Table 2 shows approximate maximum spindle speeds for different diameters of various different metals, using HSS tools, and making roughing cuts; these speeds can be increased by a factor of 3 for carbide tools, and by up to 25% for fine, light, finishing cuts. The first row, marked SFM (Surface Feet per Minute), gives the surface speed of the material at the cutting tool that the rest of the figures in the table are based on. As the diameter of the work increases, the SFM figure for a given RPM increases also; conversely, as machining progresses and the diameter being turned decreases, faster cutting speeds can be used. The figures shown are a compromise between speed of cutting and frequency of re-sharpening tool bits and can be varied accordingly.

Table 2 - Approximate spindle speeds for metal turning

Diam	Cast Iron	Stainless Steel	Mild Steel	Brass, Free Cutting Steel	Aluminium
SFM	20	40	65	117	600
1/8"	600	1200	2000	3600	18400
1/4"	300	600	1000	1800	9200
1/2"	150	300	500	900	4600
1"	75	150	250	450	2300
2"	38	75	125	225	1150

For obvious reasons, this table should be used as a rough guide only; the standard pulley set does not offer as slow a speed as 38 RPM, and the spindle bearings are rated for a maximum of 7000 RPM. There are other considerations too. If you are cutting an irregularly shaped piece of aluminium at 2" diameter, a spindle speed of 1150 RPM will generate considerable vibration, with the potential to cause the work to come free or to cause damage to the spindle bearings, so common sense and experience may well cause you to use different cutting speeds to those shown.

Carbide tipped tools often do not perform well if they are used at significantly slower speeds than those indicated in the table, so for many purposes, HSS may well prove to be a better choice when using this kind of lathe. A disadvantage of carbide tipped tools is that they are more difficult to sharpen than HSS, needing the use of diamond impregnated wheels or green (silicon carbide) grit wheels. Carbide cutting tips also tend to be less tolerant of shock loads (caused by intermittent cuts) and can be chipped relatively easily.

However, I now mostly use the "indexable" style of carbide tipped lathe tools in my Taig lathe and get excellent results. The "negative rake" indexable tools are not a good choice for a lathe of this size, though, as they require significantly more force and machine stiffness to make them cut correctly than zero rake or positive rake inserts.

Parallel turning

Much of the turning work done in a lathe will involve very simple parallel turning operations, where a length of bar stock is held in a chuck, and a right hand knife tool is used to turn from the free end of the bar up to a shoulder, or to face off the end of the bar.

If you haven't worked with a lathe before, then this is a very good place to start. Cut a short length of bar stock - say, 3/4" diameter and 3" long - and grip it in the 3-jaw or 4-jaw chuck; if the latter is used, centre it as best you can, but for this exercise, accurate centering isn't a priority. Choose a material that will be reasonably forgiving; aluminium is a good choice, and experiment with the way it behaves. Successful manual machining depends greatly on gaining a "feel" for how a material cuts; be prepared to waste a good bit of metal in the early stages, as you develop that feel. Experimenting with different depths of cut, spindle speeds, and rates of feed will help to give you an idea of what the right combination is for a given type and diameter of metal. Ideally, the cut should remove metal in a continuous ribbon; if this is not happening, then the surface finish may be impaired by the cutting tool "chattering" against the workpiece. An exception to this is brass, which tends to produce small chips (often needle sharp) rather than a continuous ribbon.

Mount a right hand knife tool in the tool post, adjusting the angle of the tool so that it can be used both to face off (moving the tool at right angles to the lathe axis, by advancing the cross slide) and to parallel turn (moving the tool in the direction of the lathe axis, by moving the saddle to the left). If the tool has been ground as shown in Figure 6, then the shaft of the cutting tool will be angled a few degrees to the left in order to achieve this, rather than being exactly at right angles to the lathe axis. When correctly adjusted, the tool will contact the workpiece only at the tip.

As with all turning operations, the turning tool should be adjusted so that it is exactly at centre height (see *Tool posts* on page 39).

Tony Jeffree

Turning between centres

Turning between centres is performed for a number of reasons:
— To turn a taper over the length of a long workpiece, by offsetting the tailstock relative to the axis of the headstock spindle.
— To turn a workpiece (either tapered or parallel) that is too long to turn without using some kind of tailstock support.
— To ensure that two or more diameters on the same workpiece can be turned, in such a way that they are all concentric with each other, even if the workpiece has to be reversed in the lathe between cuts.

Turning between centres requires the use of a dead centre in the spindle nose, and either the dead centre (on the normal tailstock ram) or the needle bearing centre to support the tail end of the work. A "drive dog" and a "catch plate" are needed to rotate the workpiece, and it is necessary to centre drill both ends of the workpiece to take the conical points of the headstock and tailstock centres. A normal centre drill is ideal for this (the kind that have a very short "pilot" drill with a larger, 60-degree drill behind it), as the hole made by the pilot drill acts as a reservoir for lubricating the centre.

There is no headstock dead centre available in the Taig accessories, so a bit of improvisation is necessary. There are at least three simple solutions to this problem:
— Use a drill chuck arbor to create a centre. Fit the arbor to the headstock and machine a 60-degree cone on the end, with the lathe tool held in the compound slide. This is a slightly expensive way to make a centre, particularly as, for accurate work, it is desirable to machine the taper "in situ" before each use. An alternative is to drill and tap the end of the arbor to take a 1/4" thread, and fit removable screw-in centres. This has the advantage of allowing the arbor to still be used for its original purpose.
— Drill and tap a blank collet to take a 1/4" thread. Make up a drawbar that will screw part-way into the back of the collet, leaving room for a removable centre to be screwed into the front of the collet. This will allow the centre to be replaced when re-machining is no longer possible.
— Grip a piece of round stock in the 3- or 4-jaw chuck and turn a 60-degree taper on the end of it. This last method has the advantage of simplicity, and also provides the catch plate "for free" (see below).

As the headstock centre will turn with the work, the centres can be made from mild steel rod, and should not be hardened.

Whatever method is used to make the centre, it will be necessary to provide a means of driving the workpiece. This is usually achieved by means of a lathe dog

(#1034; see Figure 51) and a catch plate; the lathe dog is a clamp that fits over the workpiece, and has a peg (a cap head screw) protruding from it that engages with the catch plate or one of the jaws of the lathe chuck. The Taig face plate can be converted into a catch plate simply by drilling a hole in its face that will accept the head of the cap screw, or even by bolting a small angle plate onto the face plate using one of the T-slots such that the angle plate engages with the cap screw. The cap screw should be wired to the catch plate so that it cannot move independently.

Figure 53 - Lathe dog

Figure 53 shows the Taig lathe dog being used with a centre held in the 4-jaw chuck; in this case, the lathe dog carries the driving peg (the UNF 10-32 screw at the top of the photo) which engages with the chuck jaw, acting as the catch plate as well as carrying the centre.

The choice of whether to use a live (needle bearing) centre or a dead centre in the tailstock depends on a couple of factors:

— The dead centre on the tailstock ram is not hardened, so must be used with care, slow speeds, and lots of lubrication, to avoid seizing up to the workpiece and/or scoring the centre.

— The live centre rotates with the work, so doesn't suffer the lubrication problem and can be run at higher speeds; however, as there is a bearing involved, any play or runout in the bearing will inevitably affect the accuracy of the turned part.

If the Taig needle bearing centre (#1151, Figure 38) is used, it takes the place of the normal tailstock ram, which is removed by taking out the split pin and screw/nut that attach it to the tailstock operating lever. The tailstock is brought as close to the workpiece as possible while still allowing room to do the machining necessary. The idea here is to keep the tailstock support as rigid as possible by minimising the "overhang" - the distance between the tailstock clamp and the end of the tailstock centre. The needle bearing centre is fitted in place and clamped firmly in the tailstock so that the spring-operated plunger that carries the rotating taper is fully depressed into the body of the centre. Failing to fully depress the centre fully will lead to play in the tailstock support, defeating the object of using it in the first place.

If you are planning to do parallel turning "between centres", it is important to ensure that the tailstock ram is on the same axis as the headstock (see *Checking alignment* on page 78), otherwise, the turned part will be tapered, not parallel. As observed earlier, front-to-back misalignment is much more serious in this regard than the ram being too high or too low by the same amount.

Taper turning

One of the major uses for "between centres" turning is for turning tapers by offsetting the tailstock from the centre axis of the lathe; in fact, when using an offset tailstock, between centres is the only way that any turning can be done at all. The alternative method of taper turning in the Taig is to use the compound slide (see Figure 42 on page 58, which shows the compound slide in use to cut a taper).

The ability to offset the tailstock by known amounts is very useful when taper turning, as the offset distance, plus the distance between centres, plus a bit of trigonometry, will allow a precise angle of taper to be produced. *Micrometer set-over adjuster* on page 92 describes an enhancement to the tailstock that allows the tailstock offset to be precisely controlled. The principle used to determine the necessary offset is straightforward; if the distance between the two centres is known, and the amount of the tailstock offset (from the spindle axis) is also known, then the sine of the angle of taper can be determined by dividing the offset by the distance between centres. Conversely, if you know what angle of taper you want, then the desired offset is determined by multiplying the sine of that angle by the distance between the centres. If the tailstock has previously been accurately centered, using one of the techniques described in *Checking alignment*, then the required offset can simply be "dialed in" on the micrometer adjuster.

Milling

If you have the milling slide (see Figure 39), then it is possible to perform simple, and small, milling operations in the lathe.

Milling can be considered to be the inverse of turning. In turning, the cutter moves linearly while the work is rotated; in milling, the cutter is rotated while the work is moved linearly. Milling can therefore be used to cut slots, or flat areas, or pockets in the workpiece.

The milling cutter is held in the lathe spindle; it should always be held in a suitably sized collet, and NOT in a Jacobs chuck or one of the larger chucks. There are a number of reasons for this, of which the main ones are:

— Collets provide a much more secure grip than the other types of chuck are capable of, particularly when gripping a hardened steel shank such as a milling cutter.

— The Jacobs chucks, and other self-centering chucks, tend not to be terribly accurate; the milling cutter will therefore not be as accurately centered as it should be for successful milling.

— Collets give significantly less "overhang" than any of the other chucks (i.e., the working end of the cutter is much nearer to the spindle bearings), leading to improved stiffness of the setup.

— If you are milling at the highest spindle speeds, having a 3- or 4-jaw chuck whirling away close to the workpiece adds unnecessarily to the hazards of the operation, and may partially obstruct the operator's view of the work.

The basic setup for milling in the lathe is shown in Figure 54. The milling slide is attached to the cross slide if the lathe via its two T-bolts, and the milling vice is attached to the bed of the milling slide, again by a pair of T-bolts. The workpiece (in this case, a piece of flat bar stock) is held in the jaws of the vice. There is a "register" plate at the back of the milling slide, held in place with a pair of socket head screws. The intent of this is to make it easy to mount the slide square to the cross slide; the register plate is clamped so that it protrudes below the base of the milling slide, and engages with the side of the lathe cross slide.

Figure 54 - Milling

This milling vice is intended for light duty use; the jaws are tightened by means of a thumbscrew. Do not be tempted to use pliers, etc., to tighten the screw more than hand tight, as this will simply distort the vice. If you need a stronger vice than this to do your milling, please consider the possibility that you may be in danger of overloading the machine!

The milling cutters sold as standard Taig accessories are double-ended cutters of the type known as "slot drills" in the UK; terminology varies in other countries, but the important point is that these cutters are capable of cutting when they are "plunged" into the work, much as a drill would do, whereas a more conventional end mill needs to be fed in from the side of the work. They are, as the name implies, suitable for cutting slots equal in width to the cutter diameter. If you are using cutters from other sources, the difference can be seen by examining the end of the cutter; if one of the cutting edges (there will generally be 2, 3, or 4 edges) extends across the centre of the cutter, then it can be plunged into the work. If there appears to be a central recess, with the cutting edges arranged around it, then it will not be possible to do this, and the cutter can only be fed in from the side.

The difference can be important if, for example, it is necessary to mill out a central "pocket", leaving continuous material all around the outside. If you don't have a centre-cutting end mill or slot drill, then you would need to drill a starter hole to the necessary depth in order to make it possible to use a conventional end mill. However, as cutter grinding techniques have improved with the advent of CNC, most modern end mills are capable of plunge cutting.

Table 3 - Approximate spindle speeds for milling

Cutter Diam	Cast Iron	Stainless Steel	Mild Steel	Brass, Free Cutting Steel	Aluminium
1/16"	1200	2400	4000	7200	36800
1/8"	600	1200	2000	3600	18400
3/32"	450	900	1500	2700	13800
1/4"	300	600	1000	1800	9200

5/32"	240	480	800	1440	7360
3/16"	200	400	666	1200	6133

As with turning, it is a good plan to develop a "feel" for what can be done with the milling capabilities of the lathe. A few points that are worth bearing in mind:

— A good starting point for determining the cutter speed for milling is to use the cutting speeds shown in Table 3, which should be reasonable for the 2-flute cutters sold by Taig. If 3 of 5-flute cutters are used, these speeds should be reduced by between 30% and 50%.

— When milling, always lock any slide that you are not using. For example, if the cut involves a down feed on the milling slide, the carriage should be locked using the thumbscrew at the far side of the carriage, and the cross slide should be locked by tightening the central gib adjusting screw.

— With milling, you need to take notice of the direction of rotation of the cutter and choose the direction of movement accordingly. Avoid "climb milling", i.e., moving the workpiece in the same direction that the cutting forces are trying to move the work; this tends to put more strain on the machine than conventional milling, and if the setup is not very rigid, there is a tendency for the cutter to "dig in" with nasty consequences for it and the workpiece.

— When you are milling a slot in the workpiece at the diameter of the end mill, do not attempt to cut to a depth greater than 1/2 the cutter diameter. If the final dimensions call for a greater depth of cut, achieve it in multiple passes.

— When "profiling" with an end mill (i.e., cutting in from the edge of a workpiece), the maximum depth of cut should be no more than the cutter diameter, and the width of cut no more than 1/4 of the cutter diameter. Again, if necessary, use multiple passes.

— Avoid taking cuts that are so fine that the cutter just rubs against the metal rather than cutting properly; this can prematurely blunt the cutter, and with some metals, simply work-hardens the surface, making subsequent cutting more difficult.

— Lubrication is important, both for reducing cutter wear and for improving surface finish. A simple alternative to a proper coolant system is to brush the lubricant onto the work with a small paintbrush.

Tony Jeffree

Power feed

If you have a version of the lathe that was factory-fitted with the power feed components (#M1015, for example) or if you have chosen to retrofit the power feed component kit (#1016), then the carriage can be moved either using manual control or under motor control, or both. A lathe with power feed can be seen in Figure 55.

Figure 55 - Lathe with power feed

In a power feed lathe, the rack is replaced with a leadscrew, seen below the carriage and extending the length of the lathe bed. The handwheel pinion engages with the leadscrew in the same way that it does with the rack; if the power feed is not engaged, rotating the handwheel causes the carriage to move left and right as normal, although the gearing is a factor of 3 higher than with the standard rack and pinion, resulting in three times the movement of the carriage for a given movement of the handwheel.

A small gearbox at the left-hand end of the lathe allows the leadscrew to be powered from the lathe spindle; a normal elastic (rubber) band is all that is needed to link the spindle pulley to the stepped pulley on the input shaft to the gearbox. If you fit the band to the left-most groove of the two pulleys, this will give the finest feed rate; moving the band one or two grooves to the right gives coarser feed rates. Needless to say, if you are using the power feed, then one of the spindle speeds of the lathe cannot be used at the same time, but this is a small inconvenience.

If the lathe motor is turned on, the lathe spindle rotates, which in turn powers the leadscrew gearbox, and the leadscrew starts to rotate, which also causes the carriage handwheel to rotate, but the carriage stays still. So far, so not very useful, you may say. However, the magic happens when you grip the handwheel to stop it rotating; at that point, the carriage starts moving to the left (or to the right if the drive band is twisted). Releasing the handwheel stops the carriage movement. If desired, you can increase the speed of movement by rotating the handwheel as normal, either to the left of the right. A bit of playing and you soon get the hang of it; the result is a great improvement in the ease of use and versatility of the lathe.

It should be noted that this power feed arrangement is not capable of being used for thread cutting operations and should not be considered as a replacement for a conventional leadscrew and change wheels for that purpose.

Chapter Seven
Checking alignment

The Taig lathe is manufactured such that there should be little need for adjustment of the basic alignment of the lathe. However, it is always good to be able to convince yourself that the lathe is correctly aligned, and to adjust the alignment if that proves necessary. An invaluable aid to checking alignment is a dial indicator. This is a device that looks rather like a stopwatch but has a spring-loaded plunger protruding from the bottom; as the plunger is depressed, the deflection shows up as movement of the hands of the dial indicator. The dial indicator must be held firmly relative to the thing that is being measured; small adjustable stands with magnetic bases can be used for this purpose, as can purpose-built fixtures that allow it to be held in a tool post or in one of the lathe's T-slots. Examples of how to use this device will be found in the photographs that illustrate this section.

In some cases, although it may be possible to detect misalignment, there may be nothing that can easily be done to correct the problem. However, this need not prevent the lathe from being used to produce accurate results, if the nature of the problem is known, and appropriate machining techniques are used to circumvent the problem.

Spindle runout

For obvious reasons, it is important that the lathe spindle is concentric with its axis of rotation. The extent to which the business end of the spindle deviates from its axis of rotation is known as runout. This can be measured as illustrated in Figure 56.

Figure 56 - Measuring spindle runout

The tip of the dial indicator's plunger is adjusted so that it contacts the "register" of the lathe spindle, the cylindrical area between the threaded section of the spindle and the hex-section shoulder. By rotating the lathe spindle by hand and observing any deflection on the dial, the degree to which the register is not concentric with the axis of rotation can be determined. On the author's lathe, the dial indicator showed about half a thou of total deviation (subtracting the minimum reading from the maximum reading on the dial); in other words, there is about 0.00025" of offset between the centre of the spindle and its axis of rotation. As this is well within the 0.0004" quoted as the maximum bearing runout for the lathe, this is clearly OK.

If the measured runout is excessively large, then this indicates that there is something significantly wrong with the spindle and/or its bearings. If the latter, then there may be detectable play in the bearings, or noise or "grittiness" when the spindle is rotated. Other reasons for excessive runout are damage to the spindle itself, perhaps if it has been bent as a result of some accident or excessive load.

In any event, there is not much that can be done about excessive spindle runout other than replacing the bearings; in reality, as the bearings are a press fit on the spindle, it is simpler to replace the whole spindle and bearing assembly, known as a *spindle cartridge*.

Axial alignment

For the lathe to be able to cut a true cylinder, the axis of rotation of the spindle has to be parallel to the axis of the lathe bed. This strictly must be true in both the horizontal and the vertical plane; however, misalignment in the vertical plane is far less significant than in the horizontal plane in terms of its effect on the finished product.

For example, consider what happens if the spindle axis diverges from the lathe bed axis by 5 thou in the horizontal plane (i.e., measuring from the near edge of the lathe bed to the spindle axis would give 5 thou greater measurement at the tailstock end than at the headstock end; the spindle is pointing slightly away from the operator). If the operator tried to machine a long piece of round bar stock, then what the operator would intend to be a straight cut over the length of the lathe bed would result in the workpiece being 10 thou fatter at the tail end than at the headstock end—a cone rather than a cylinder.

In contrast, consider what happens in the same situation if the spindle axis diverges from the lathe bed axis by 5 thou in the vertical plane (i.e., measuring from the top surface of the lathe bed to the spindle axis would give 5 thou greater measurement at the tailstock end than at the headstock end; the spindle axis is pointing slightly up towards the ceiling). If the lathe tool is set accurately on centre height at the centre of the lathe bed, then at the tailstock end it would be 2.5 thou below the centre of the workpiece, and at the headstock end, 2.5 thou above the centre of the workpiece. So, if the cutting tool was adjusted so that it was just in contact at the middle of the workpiece, then at either end of the workpiece, the curvature of the bar stock will mean that the tool tip is not in contact with the surface of the bar, but by considerably less than 2.5 thou. Making what the operator would intend to be a straight cut over the length of the lathe bed would result in the workpiece being slightly fatter at either end than it is in the middle; the resultant surface will be very slightly "waisted". However, the difference in diameter between the ends of the bar and the middle will be very small. For example, for a 1" diameter cylinder machined with the spindle misaligned by this amount, the ends would be larger than the middle by only 0.0000063" (6.3 millionths of an inch). Because the error depends on the curvature of the workpiece, the error increases with smaller diameters; however, even with a 1/8" diameter cylinder, the difference between the ends and the middle would still only be 0.00005".

So, in both cases, the misalignment results in something that is not the intended perfect cylinder; however, the vertical misalignment is much less of a problem in terms of its effect on the finished article.

With larger lathes, a pre-cursor to achieving correct axial alignment is to "level" the lathe bed; this is literally done by sitting a very accurate spirit level on the lathe bed, and adjusting the lathe mounts (usually constructed in the form of small screw jacks) until the bed is level at all points along its length, and in both axes. The primary reason for this has nothing at all to do with getting the lathe bed level as such but is a convenient means of ensuring that the lathe bed (which is generally a heavy iron casting) is not under any stresses that would cause it to twist, and that would therefore upset its accuracy. With the Taig lathe, which has a relatively light, cantilevered lathe bed, there is no need to level the lathe bed at all, as this has no effect on its accuracy.

Checking axial alignment quickly

It is possible to make a crude check on the state of the lathe's axial alignment with the aid of a 4-jaw chuck, a "test bar", and a dial indicator. For these purposes, the test bar is any length of round stock that you are reasonably confident is cylindrical and straight; for example, a length of "centreless-ground" bar. Silver steel rod (known as "drill rod" in the USA) is suitable for this purpose; something around 3/8" or 1/2" diameter is fine.

Grip the bar in the 4-jaw chuck and use the dial indicator to adjust it so that it is as near as possible to being on-centre. Hold the dial indicator in the tool post, and as close to the chuck as possible, to do this, as shown in Figure 57; adjust the dial indicator so that the tip of the indicator touches the bar on centre height. What you will probably notice at this point is that, as the chuck is rotated, the far end of the bar describes a (hopefully small) circle; this will reflect any inaccuracy in the geometry of the chuck jaws, any runout in the spindle itself, and possibly also the fact that the bar is not quite straight.

Figure 57 - Axial alignment 1

However, if you rotate the chuck so that the end of the bar is at the top (or bottom) of the circle (i.e., the bar is pointing as far up or down as it goes in the course of a revolution), then the side of the bar should be near enough parallel to the axis of the spindle. Moving the carriage from the left hand end of the bar to the right hand end (see Figure 58), with the dial indicator still in place, will give a rough indication of whether the spindle axis is aligned with the axis of the bed.

Figure 58 - Axial alignment 2

Using this technique, my lathe showed a thou or so difference in dial reading between the headstock and tailstock end of the bar, some of which would have been due to the change in centre height, as the bar showed significant runout (about 10 thou) over that distance, much of which would probably be correctable by trueing up the mating faces of the 4-jaw chuck jaws.

Checking axial alignment accurately

There is no substitute for actually cutting metal on the lathe, and then measuring the result, in order to see how the lathe is aligned. Clamp a length of round stock—at least 3/4" diameter, preferably larger—in the 3-jaw chuck, so that about 2" of the bar protrudes from the chuck jaws. Face off the end of the bar, and very carefully turn down the length of the bar so that the entire surface has been machined. Use a freshly sharpened HSS right hand knife tool for this. With this amount of overhang, there is a great risk of tool chattering, so take it carefully. Using a free cutting "leaded" steel, or a clean cutting aluminium, helps avoid this problem.

Hone the cutting tool once more to optimum sharpness, and make a final, very fine cut by advancing the tool by as little as you can. Aim for well below a thou of cut, and make several very slow passes over the surface, using a high spindle speed, to ensure that any effect caused by the tool deflecting the workpiece is minimized. The result will be as shown in Figure 59.

Figure 59 - Axial alignment 3

Using a micrometer, carefully measure the diameter of the finished cylinder, at each end. Any difference in the diameter is (as per the earlier discussion) a reflection of any misalignment that is present, and most probably, of misalignment in the horizontal plane.

The bar shown in Figure 59 measured 0.5772" at the chuck end, and 0.5775" at the free end, giving a difference of 0.0003", equating to a misalignment in the horizontal plane of 0.00015" over 2 inches, or roughly 0.00075" over the usable length of the lathe bed. This is small enough to be insignificant for most purposes.

If this test shows a significant misalignment, then this could be due to a number of things:

- Over-tightening of the socket-head clamping screw(s) at the back of the headstock that clamp it to the lathe bed[11]. If the screws are tightened too enthusiastically, the aluminium extrusion that forms the headstock shell can distort. Loosen these off, ensure that the headstock is properly engaged with the lathe bed dovetails, and then re-tighten with more moderate torque on the hex wrench.
- Dirt or swarf trapped in the dovetails. Remove the headstock from the lathe bed, thoroughly clean the mating surfaces of the lathe bed and the headstock shell, and re-fit.
- On the old-style headstock, dirt or swarf trapped between the outer race of the spindle bearings and the two headstock shell extrusions. Disassemble the headstock (one countersunk socket-head screw each side), clean the components and reassemble.
- Far less likely than any of the above, there could be an inherent misalignment due to faulty manufacture of the lathe bed and/or headstock, or these components could have been damaged in some way. In either case, replace the faulty components.

Adjusting the tailstock - quickly

When drilling from the tailstock, and when using the tailstock to support the workpiece, it is necessary to adjust the tailstock such that the axis of the ram coincides as closely as possible with the axis of the lathe spindle. The exception to this is, of course, when you are taper turning "between centres". The simplest way of adjusting the tailstock to align it with the spindle axis is to align the point on the end of the tailstock ram with a centre held in the lathe chuck, as illustrated in Figure 60.

Figure 60 - Tailstock alignment

---

[11] The old-style headstock, that consisted of a pair of "clam shells" held together with a pair of socket head screws, was more prone to such distortion problems.

In the right-hand picture, there is a visible misalignment; in the left-hand picture, the tailstock is much more closely aligned with the centre. The adjustment is achieved by loosening the socket head screw at the back of the tailstock and sliding the tailstock column on its dovetails; this is very much easier to do if you have made the micrometer set-over adjuster described in *Micrometer set-over adjuster* on page 92. It is surprising how accurately this can be done just by eye alone; with care, and with the assistance of a magnifying glass if necessary, considerable accuracy can be attained this way.

However, for it to be truly accurate, the centre held in the chuck must itself be accurate; this can only be ensured by re-cutting the point every time it is used, as mentioned in the discussion of turning between centres. Figure 61 shows the centre being cleaned up, using the compound slide adjusted to the angle of the centre.

Figure 61 – Trueing a dead centre

Unfortunately, using this technique on my own lathe tailstock demonstrates quite clearly that the tailstock ram is too low by a noticeable amount (see Figure 62). The only way to deal with this is to adjust the height of the tailstock, by reducing its height if the ram is too high, or, as in this case, where the ram is too low, by increasing its height.

In reality, the amount of misalignment represented by Figure 62 is not a serious problem for most purposes; it is not enough to worry about when drilling from the tailstock, and when turning between centres, as already discussed, vertical misalignment is far less significant than it would be if it was in the horizontal plane. However, if you discover this kind of problem and really want to fix it, then it can be done.

Figure 62 - Tailstock vertical misalignment

Increasing the tailstock height is rather easier to achieve than decreasing its height, as it can be done by adding shims of a suitable thickness between the base of the tailstock and the top surface of the lathe bed. These can be fixed in place under the tailstock with a drop of Superglue to make a reasonably permanent height adjustment. However, it is also possible to reduce its height, by shaving, milling, or scraping off a suitable amount of material from the tailstock components. Probably the easiest place to do this is at the base of the top half of the tailstock, where there are two pads at the bottom of the dovetail that bear onto the upper surface of the bottom half of the tailstock. However, this should only be attempted with care, as it would be easy to create further problems by taking different amounts off from front and back, for example, thereby causing the ram to point up or down rather than be horizontal.

In either case, the first step has to be to accurately determine the thickness of shim to add, or thickness of metal to remove; this is the subject of the next section.

Checking tailstock alignment - accurately

As before, accurate measurement calls for the use of a dial indicator; this time, the idea is to keep the ram steady and rotate the indicator. A useful setup for this is shown in Figure 63; here, the face plate has been fitted to the spindle nose, and the dial indicator is held in place with a T-bolt fixed into one of the face plate's T-slots, the other end of the T-bolt serving as a set screw for a brass collar that holds the indicator in position. The piston of the indicator bears against the body of the tailstock ram, at right angles to the axis of the ram, adjusted so that the piston points along a radius at the central axis of the ram.

THE TAIG/PEATOL LATHE

Figure 63 - Measuring tailstock alignment

As the spindle with attached face plate and dial indicator is rotated, any deflection of the indicator shows by how much the ram is off-axis. It is a little difficult to read the dial in the orientation shown; rather better to rotate the dial to the vertical, but this way gives a clearer illustration of what is going on.

To set the horizontal alignment, simply note the readings between the front and the back of the ram (the two positions where the indicator piston is horizontal), and adjust till there is no deflection, as before. Again, the micrometer adjuster mentioned in *Micrometer set-over adjuster* on page 92 makes this easy to do. Having set the horizontal alignment, the difference in the indicated reading between the horizontal and vertical positions tells you by how much the ram is misaligned vertically (too high or too low). On my lathe, this showed that the ram is almost exactly 3 though too low; hence, adding 3 thou of shims under the tailstock will correct the problem. However, as the consequences of such a small misalignment are minor, it may be a while till I get around to fixing it.

Chapter Eight
Enhancing the lathe

This chapter looks at some ways that the lathe can be enhanced, and also at some simple accessories that increase the lathe's usability and versatility.

Spinning handles

The handles fitted to the cross slide dial and the saddle traverse handwheel are simple brass handles that have been slightly "waisted", but do not rotate as the dials are turned. While they very quickly become polished through use and slip through the fingers as the dial is turned, I have a preference for handles that turn independently of the dial—I find them more comfortable to use, and easier to maintain a constant feed rate.

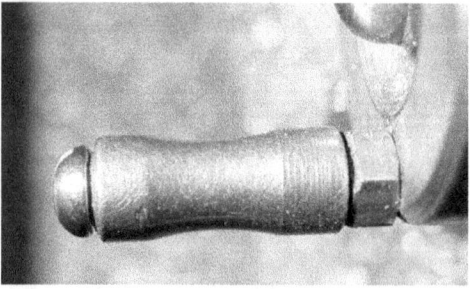

Figure 64 - A spinning handle

The existing handles are made of brass and are a press-fit in the respective dials; it is easy to remove them without damage with careful use of flat-faced pliers.

Having removed a handle, it is held in the 3-jaw chuck and drilled axially to take a suitable diameter bolt, which will act as the spindle for the handle. I have found that dome-headed brass 4BA bolts are perfect for this purpose, as the holes in the dials into which the handles were a press-fit will take a 4BA tap without further drilling. If 4BA is used, then a 3.6mm axial hole in the handle will allow the handle to spin freely on the bolt.

A matching brass nut is run onto the end of the bolt and fixed in place using Superglue or Locktite so that the handle is still free to spin but has minimal axial play on the bolt. The hole in the handle is tapped to suit the chosen bolt, and the modified handle screwed in place, again with a drop of Locktite to prevent it coming loose. The end result can be seen in Figure 64.

Locking knobs and handles

There are several places on the Taig lathe where cap head screws are used for locking and adjusting various components. Many of these are adjusted regularly as the lathe is being used, and constantly reaching for the right hex wrench can be irritating. It is a simple job to replace or modify these screws so that they have their own handles.

The main candidates for this treatment are:
— The three locking screws on the tailstock (one that locks the tailstock to the carriage, one that locks the set-over adjustment slide, and one that locks the tailstock ram).
— The T-bolts used in the tool posts.
— The compound slide mounting/angle adjusting screw.

Replacing the tailstock screws with knurled knobs is a straightforward job. The screws are UNF 10-32 and about 1" long, and replacements can be turned from 5/16" mild steel round stock. The overall length of the replacement screw should be about 1.75"; turn down a 1" length of the stock to the right diameter for threading 10-32 (0.19") and thread it with a UNF 10-32 die. I find that threading small screws like this is easily done in the lathe, using a 1/2" chuck in the tailstock pushed up against the back of a hand die holder to keep it square, and turning the lathe chuck by hand. Of course, if you have invested in the tailstock die holder (Figure 47), this operation can be even easier!

Figure 65 - Tailstock locking knobs

The full length of the screw will need to be threaded. The screw can then be parted off from the parent stock, leaving a 3/4" length of 5/16" diameter material as the "head" of the screw. A knob can then be fashioned from round bar of a suitable diameter—ideally about 1.25" diameter—bored 5/16" for the screw head and parted off to 1/2" thick after knurling the outer surface to improve grip. The screw is then Superglued into the hole in the knob after cleaning off any surface oils and swarf. The result can be seen in Figure 65.

The mounting/adjusting screw for the compound slide can benefit considerably from a similar treatment, as fitting the slide involves using an over-length hexagonal wrench inserted along one of the cross slide T-slots, which is very fiddly. In this case, the overall length of the screw needs to be about 3", with the threaded portion taking up the last 1 1/8", threaded UNC 6-32. Round stock of 7/32" diameter is about right for this screw; again, a knurled knob is fitted to the thick end.

Extending the tailstock lever

The standard tailstock lever is serviceable but can be on the short side for drilling larger holes. It can also be rather uncomfortable to use, as it is basically just a length of 1/2" X 1/8" steel strip, with square corners that can dig into the hand. Extending the lever to about twice its length, and at the same time, giving it a more comfortable handle, considerably improves the usability of the tailstock for drilling operations.

Figure 66 – Extended tailstock lever

Figure 66 shows an example of how this can be achieved—it took about half an hour to make this handle, using some materials from the scrap box. The outer portion of the handle was made from a 6" length of 15mm OD copper plumbing pipe; a blanking cap (also a standard plumbing fitment) soldered on one end nicely finishes off the free end of the handle. The inside diameter of this pipe is a little larger than the 1/2" width of the tailstock lever, so I cut a 2" length of 1/2" diameter brass rod, and cut a 1.5" long, 1/8" wide, axial slot in the rod to accept the lever. This slotted brass rod was then soft soldered into the open end of the tube, leaving the slotted end flush with the end of the tube.

Cutting the slot can be achieved with care by using a hacksaw; make two longitudinal cuts straddling the axis of the rod, and then cross-cut at the base of the cuts with a piercing saw to take out the sliver of brass that is left between the two long cuts. The handle is then fitted onto the tailstock lever, and two holes drilled for fixing screws. These holes should first be drilled right through the tube, brass insert, and lever, using a tapping drill—I used 4BA screws, and a 3.1mm tapping drill. Then the holes are opened out through the tube, the first half of the insert, and the lever, leaving the hole through the second half of the insert still at tapping diameter. The hole is then tapped, and the screws fitted. Tightening the screws has the effect of clamping the new handle onto the existing lever.

Having made this modification, the leverage obtainable with the tailstock is considerably increased; do not be tempted to apply more force to the lever than is appropriate for the machining operation being performed, as damage to the lathe and the workpiece may result. If you find, for example, that a drill being fed from the tailstock isn't cutting terribly well, maybe the real reason is that the drill isn't very sharp! Of course, as the handle is made from copper tube, there is a natural limit to the force that can be applied before the tube bends, which is probably just as well.

Micrometer set-over adjuster

The tailstock can be adjusted for taper turning, as described in *The lever operated drilling tailstock* on page 54. If taper turning using tailstock offset is something that you are likely to want to do fairly often, it is useful to be able to offset the tailstock by controlled amounts, and to be able to return the tailstock to the lathe axis again when the taper turning is complete.

Figure 67 - Micrometer set-over adjuster

Figure 67 shows a modification to the tailstock to achieve this end. The micrometer adjuster is rather similar in construction to the handwheel and leadscrew that advances the cross slide; both use a 1/4" 20 TPI leadscrew and a handwheel with 50 graduations, giving adjustment in thousandths of an inch. The left-hand photo shows the handwheel; the right-hand photo shows the "lead nut"—

a 3/8" thick steel plate that is attached to the base of the tailstock with a pair of socket head UNF 10-32 screws. In this prototype, I have chosen to use countersunk screw these into the base itself; a slightly easier option is to use the T-slots provided on the rear of the tailstock base, and this is the approach used in the following description. The photo of the handwheel shows a hole where a handle was originally fitted; however, as I found that it fouled my newly extended lever, and proved to be unnecessary anyway, I removed it; hence, this does not feature in the description below.

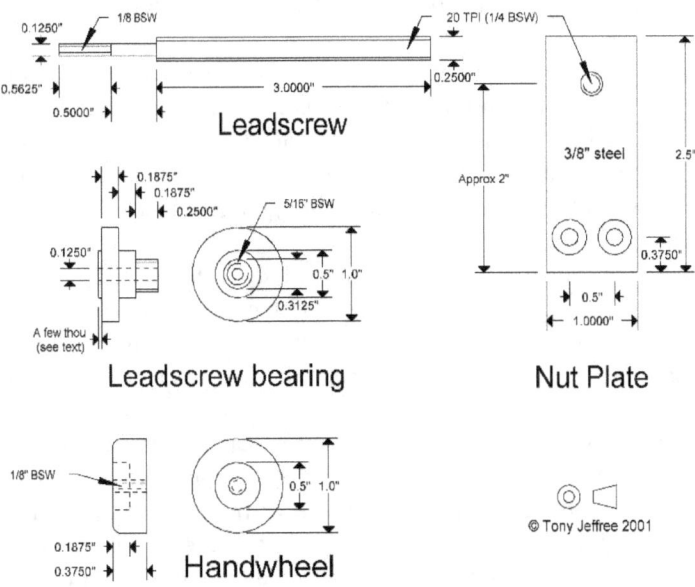

Figure 68 - Micrometer set-over adjuster parts

Figure 68 shows the main parts involved in making the adjuster. The handwheel is cut from 1" diameter material - brass, aluminium or steel, as you see fit. The first job is to face off the stock and bore the 1/2" diameter recess in the face of the handwheel and give a slight radius to the edge of the wheel. The tapping hole for the 1/8" BSW thread is then drilled and tapped. The handwheel can then be parted off to 3/8" thick. I engraved my handwheel with 50 graduations (one thou movement per graduation) using my CNC mill; for those without this very useful facility, engraving the graduations will involve some other means of indexing, such as those discussed in *Dividing and graduating* on page 107.

The leadscrew bearing is also cut from 1" diameter stock; again, the choice of material is not critical. This is faced off, drilled and reamed 1/8", and the two large steps shown in the diagram are machined. The first reduces the diameter to 5/16" over 1/4", which is subsequently threaded 5/16" BSW; this thread will be used to lock the bearing into the body of the tailstock. The diameter of the second step is not critical. The bearing can then be parted off from the parent stock and reversed in the chuck to true up the face. Note that the central 1/2" diameter of this face protrudes a few thou; this forms the bearing surface that the handwheel will bear against.

The leadscrew itself is cut from a length of 1/4" 20 TPI threaded rod. This can be held in the 3-jaw chuck to machine down to 1/8" diameter and thread the last 1/2" or so 1/8" BSW to match the handwheel. You will also need a small jam nut tapped 1/8" BSW; I made a domed nut from a 1/4" length of 1/4" diameter brass rod, with two flats filed on it.

The leadscrew and bearing assembly can now be assembled. The thin end of the leadscrew fits through the thin end of the bearing block, and screws into the back (non-recessed) face of the handwheel. This should be adjusted for minimum play consistent with the bearing still being able to rotate. The jam nut is then threaded in place with a drop of Locktite thread locker and tightened to lock the handwheel onto the leadscrew.

The next step is to cross-drill the tailstock body with a 5/16" BSW tapping drill. The hole passes right the way through the tailstock, in the middle of the thinnest part of the tailstock body; this is approximately 2" above the bottom of the complete tailstock assembly. The near end of this hole can then be tapped 5/16" BSW to take the leadscrew bearing.

The last item to make is the nut plate. This should be fairly substantial - I used 3/8" mild steel, and the diagram is fairly self-explanatory. The plate should be held in place with its T-bolts so that the position of the 1/4" 20 TPI can be spotted through from the hole in the tailstock body; the hole is then drilled and tapped.

Finally, the leadscrew assembly is screwed into the tailstock body, again using a drop of Locktite to lock the threads, and the leadscrew threaded into the nut plate.

Setting the tailstock on-axis is made much simpler and quicker with this device. For "quick-and-dirty" adjusting, where high accuracy is not required, for example, when drilling from the tailstock, proceed as follows. A piece of soft material (brass or aluminium) is held in the three-jaw chuck and faced off. The tailstock is brought up to the chuck, locked in position, and the tip of the ram used to carefully score a circle on the end of the cleaned-up face of the soft metal. Do this by turning the chuck by hand, rather than under power. The radius of the scored circle gives the distance that the tailstock is offset from the lathe axis.

For more accurate uses, for example when using the tailstock for turning between centres, the most accurate way of setting the tailstock is to turn a test piece between centres and measure its radius at its extreme ends. The difference in radius gives the offset that is required. Alternatively, a dial indicator held in the headstock chuck, adjusted so that it indicates against the circumference of the tailstock ram can be used to determine how off-centre the tailstock is (see Checking alignment on page 78); however, this will not take into account any lack of concentricity that may exist between the pointed end of the tailstock ram and the ram itself.

A filing rest

Figure 69 - Filing rest

It is often useful to be able to accurately file flats on a workpiece held in the lathe; for example, filing the winding arbor for a clock to produce a square section at the end of the arbor to fit the winding key. A filing rest provides a useful means of achieving this end; in essence, it consists of a pair of metal rollers that can be positioned either side of the workpiece and can be raised or lowered to the desired height. The file is able to cut down into the workpiece until it contacts on both rollers, at which point it can cut no deeper. The workpiece can then be rotated and locked in the position for the next cut, and the process is repeated.

Figure 69 shows a filing rest suitable for use with the Taig lathe; the design shown here is an adaption of one shown in John Wilding's book "Using The Small Lathe", designed to suit a small Toyo lathe. The base of the filing rest is cut from one of the Taig riser blocks (see Figure 44); if you have a pair of these blocks, you will find that the tailstock rising block is just under an inch longer than is necessary to support the tailstock. The excess 7/8" or so of the rising block can simply be sawn off, complete with its clamping screw, and the cut ends filed smooth; the shorter piece makes the perfect base for this project. It will then be necessary to fit a second clamping screw to the shortened tailstock riser block to restore it to full usability.

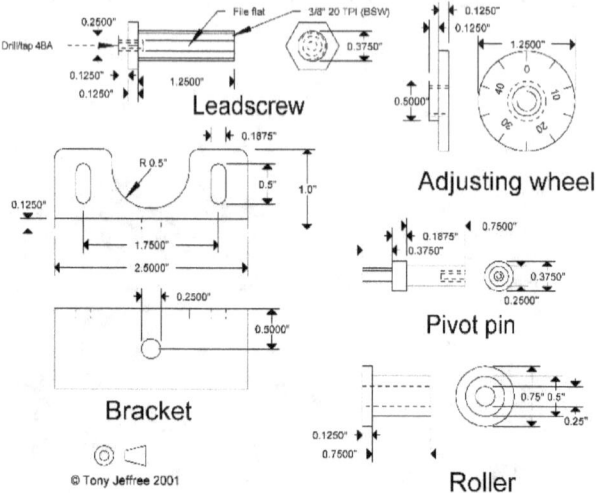

Figure 70 - Filing rest parts

Figure 70 shows the main components of the filing rest. The leadscrew is machined from a 3/8" BSF (20 TPI) hex head bolt, or from a length of 3/8" BSF studding with a nut soft soldered or Locktited onto one end. The threaded end is held in the 3-jaw chuck to machine the head, reducing its diameter to 1/4" over a 1/8" length. While the screw is still held in the 3-jaw, an axial hole is drilled and tapped 4BA; this will be used to attach the screw to the rest. A flat is then filed on the entire length of the threaded portion of the screw.

The adjusting wheel can be machined in one piece from 1.25" diameter brass stock, drilled and tapped 3/4" BSF. A full turn of this adjusting wheel on the leadscrew advances the wheel (and therefore, will adjust the height of the rest) by 50 thou. The wheel should ideally be graduated to suit; ten graduations, indicating 5 thou increments, is about right. See *Dividing and graduating* on page 107 for further reading on tools and techniques for achieving this.

The bracket is made from a 1.75" length of 1" X 1" X 1/8" aluminium, brass or steel angle; the profile can be achieved by hacksawing and filing after marking out using "engineers' layout blue" and a scriber. A useful alternative to using layout blue is to coat the surface with a waterproof marking pen. The two vertical slots in the bracket allow adjustment of the position of the rollers; these can be formed by drilling the extreme ends of the slots with a 3.16" drill and then filing out the rest, or by milling using the vertical milling slide (see Figure 39). The hole to take the end of the leadscrew is drilled 1/4" diameter.

The base should be drilled vertically with a drill that is as close as possible to the outer diameter of the leadscrew. In the case of the sample of 3/8" BSF studding that I used, a suitable drill size was found to be 9.4mm. The vertical hole is placed half-way between the dovetails, so that when the rest is fitted to the base, the rollers will be positioned either side of the lathe spindle axis. Cross drill the hole in the base to take a UNF 10-32 set screw—this hole should be placed about 1/4" below the top surface of the base. The bracket can then be attached to the leadscrew with a 4BA screw, orienting the leadscrew so that, with the flat facing the setscrew in the base, the bracket will sit at right angles to the lathe axis, as seen in the photo. I machined a small brass washer to sit under the screw head, made from 3/8" diameter brass rod, drilled and countersunk to fit the screw, and parted off 3/32" thick. Two more of these washers are needed to hold the rollers in place—see later.

Two pivot pins and two rollers are made from mild steel round stock, as per the diagrams. The rollers are machined from 3/4" stock, reducing the diameter to 1/2" over a 3/4" length, and leaving a further 1/8" shoulder. The axial holes in the rollers should be drilled and reamed 1/4" while still in the chuck, before parting off, so that the bores will be coaxial with the 1/2" diameter. The pivot pins should be machined so that they are a close running fit in the rollers, and so that the 1/4" diameter portion is slightly (i.e., a few thou) longer than the overall length of the rollers. The rollers are held in place on the pivot pins with two more 4BA screws and 3/8" diameter brass washers; the pivot pins in turn are held in the slots in the bracket by means of 2BA nuts and plain washers.

Using the assembled rest is fairly obvious - the shoulders on the rollers act as a guide, limiting where the file can cut, and ensuring that a clean shoulder can be filed on the part. The height of the rollers is adjusted by loosening the set screw, rotating the adjusting wheel, and re-tightening the set screw.

Machining small screws

Figure 71 - A screw held in a tapped collet

It is often desirable to modify small screws—either to clean up the heads of machine-made screws so that they look more attractive when they are used in exposed positions, or to machine the threaded end of the screw in some way.

Blank collets can readily be pressed into service for the first kind of use—fit a blank collet to the spindle nose and tighten the collet closer, then drill the collet from the tailstock with a tapping drill for the thread concerned, and tap the hole. The tailstock chuck can be used to advantage when tapping holes that have been drilled on centre—replace the tapping drill with the tap, loosen the clamping screw so that the ram is free to move, advance the tap into the hole and turn the lathe spindle by hand (first making sure the power is switched off). Tapping in this way helps ensure that the resultant thread is on axis. Turning the lathe by hand can be a problem if you haven't got a chuck to grab hold of, so in this particular case, the lathe crank handle described on page 101 comes in handy.

Having tapped the hole, screws can be screwed into the collet with their heads protruding; it may be necessary to run a locknut onto the screw first in order to position the screw correctly to allow full access to the head. If you need to do this kind of thing often, it may be worth making up a set of these threaded collets in preferred sizes, to save time later. Figure 71 shows one of these tapped collets in use—in this case, to clean up the head of a brass 4BA cheese-headed screw.

Holding screws the other way around, so that the threaded end can be machined, can be achieved by using the collet closer, along with a blanking plate and a series of stepped washers machined to fit the hole in the collet closer and bored to the diameter of the screws. The blanking plate can again be machined from a spare blank collet - in this case, the fat end of the collet is machined off as far as the start of the taper, so that the resultant blank ends up flush with the spindle nose when it is inserted in the spindle taper.

Figure 72 - Machining a blank collet

Figure 72 shows this machining operation in progress, using the needle bearing centre to give tailstock support. It is extremely difficult to complete this machining operation without knocking the collet out of the chuck jaws a few times, even with the tailstock support, as the thin end of the blank has a very short parallel section to grab with the jaws. An easier procedure might be to axially drill and tap the collet for a draw bar to fit the rear (narrow end) of the blank collet, and then machine a suitable screw to length that can fill the threaded hole from the front of the collet. This screw can then be Superglued in place; the collet and drawbar could then be fitted to the spindle and faced off flush with the end of the spindle nose.

Figure 73 - A selection of screw head collets

Figure 74 - Machining the end of a screw

The stepped washers are machined from brass, aluminium or steel round stock; they are 0.6" in diameter at their widest, and about 5/32" thick. The step is machined to be a close fit in the central hole in the collet closer; this varies slightly between samples of collet closer, mine being 0.434" in diameter. After machining the step to a depth of about 1/8", the washer can be drilled from the tailstock with a clearance hole to match the screw that it will be used to hold. Figure 73 shows a set of five of these collets, drilled for 0, 2, 4, 6 and 8 BA screws, along with the modified blank collet and the collet closer.

Figure 74 shows the end of a 2BA screw being machined, using one of the screw head collets. Note that machining a screw of this length without tailstock support is a delicate operation and must proceed with very light cuts if you are to avoid bending the screw!

THE TAIG/PEATOL LATHE

Handwheel adapter

Sometimes, it is convenient to be able to attach accessories to the rear end of the lathe spindle. For example, it is sometimes useful to be able to turn the lathe spindle by hand, rather than under motor power—when tapping holes in the lathe using a tap held in the tailstock chuck, attempting to tap under power is a recipe for broken taps and spoilt work, whereas the use of a crank handle attached to the lathe spindle makes hole tapping a simple and straightforward job. A second example is that it is sometimes useful to attach a division plate or a spur gear to the spindle, as an aid to dividing or graduating operations in the lathe (see *Dividing and graduating* on page 107). A detent can then be used to engage with a hole in the plate or a tooth in the gear, allowing the spindle to be locked at known positions while work is carried out on the workpiece held in the chuck.

Figure 75 shows an adapter that will allow a handwheel to be used to drive the Taig lathe spindle, as seen in Figure 76. This particular handwheel is part of the Sherline threading kit used in *Adding a leadscrew* on page 145; however, a similar handwheel could be fabricated from a metal (or even timber) disc, fitted with a simple handle. This same principle could be used to hold a dividing plate or gear onto the end of the spindle.

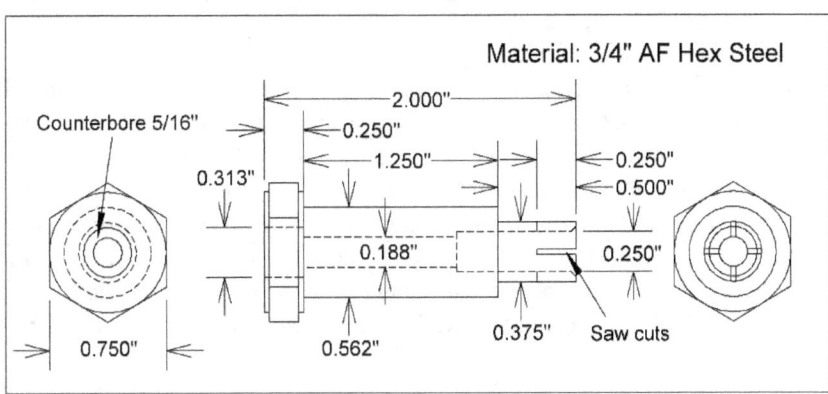

Figure 75 - Handwheel adapter

The handwheel has a 9/16" bore and is fitted with a set screw; the adapter passes through the handwheel and is fitted into the 3/8" bore of the Taig pulley. A 1.75" 10-32 UNF socket head bolt passes through the adapter, and screws into a conical nut (Figure 77) that expands the end of the adapter to grip the bore. The hexagonal head simplifies tightening the 10-32 bolt.

Figure 76 - Handwheel drive

The adapter is made from ¾" AF hexagonal steel stock. A ¼" shoulder of hexagonal material is left at the head end; the next 1.25" are turned down to 9/32" diameter, and the remaining ½" turned down to 3/8". Drill the piece axially 3/16" diameter, counterbore ¼" diameter to a depth of ¾" and add a slight internal bevel at the thin end. Part off to length, reverse and counterbore 5/16" to a depth of ¼" to take the bolt head. Remove from the lathe and make two cross cuts in the thin end to a depth of ¼" to allow the collet to be expanded by the conical nut. The latter is a simple turning job using 3/8" steel stock, tapered then drilled and tapped 10-32 UNF.

THE TAIG/PEATOL LATHE

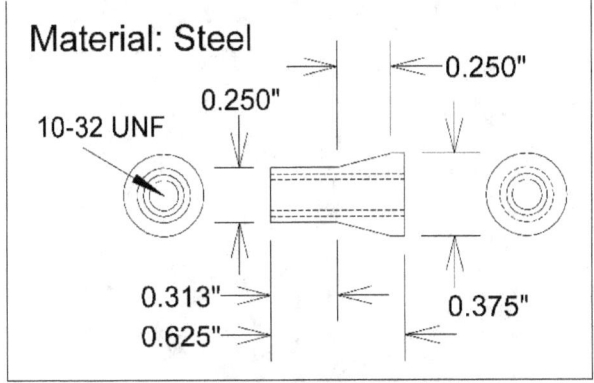

Figure 77 - Conical nut

Additional chucks

Small chucks suitable for use in a lathe of this size can easily be obtained from the usual suppliers or the second-hand market and can prove to be useful additions to the lathe's capabilities. In particular, as the Taig accessory range does not include a 3-jaw chuck with hardened jaws, it may be desirable to obtain one of these for some types of operation.

Fitting a non-Taig chuck to the Taig spindle will involve machining a suitable adapter or backplate; the most convenient type of chuck to use is one that is recessed at the back to take a backplate, although it may be possible to make up adapters for other chucks. If you are lucky enough to obtain a chuck threaded for the Taig spindle, this is of course ideal; however, check that the chuck seats properly on the "register" portion of the spindle (the part that is not threaded, just to the left of the thread), as this is important for keeping the chuck concentric with the spindle. The register diameter should closely match the diameter of the opening in the back of the chuck, otherwise, the chuck will have excessive runout.

Figure 78 – A non-Taig 3-jaw chuck

Backplates for chucks can conveniently be machined from Taig face plates (Figure 14). The face plate is mounted on the spindle, and a shallow step machined on the face such that the raised portion that is left mates with the chuck's recess with as little play as possible. The fit is important, as the recess acts as a register that will ensure that the chuck will be as concentric as possible with its backplate. It is worth allowing the backplate to cool down before taking the final cuts to bring the register to size, as thermal expansion can make a significant difference to the fit!

The periphery of the backplate is machined off to match the diameter of the body of the chuck, and then the backplate is removed from the spindle. Spot through the bolt-holes in the chuck body to mark the positions of the mounting holes in the backplate and drill the holes. Generally, the chuck body is tapped to accept the set of mounting bolts supplied with it; however, sometimes, the chuck is not threaded, and the bolts are intended to thread into the holes in the backplate, so this should be borne in mind when choosing the drill diameter for the holes in the backplate.

Figure 78 shows a small 3-jaw chuck mounted in the way I have described. This chuck is of Far Eastern origin, with a body diameter of 2.5". I found that the hardened jaws had been poorly finished, so even though the chuck was mounted with its body concentric with the backplate, items gripped in the chuck were significantly off-centre. This was remedied by careful grinding of the points of the jaws. I used a small mounted grinding wheel of the kind sold for use in mini drills such as the Dremel; this was held in the Taig tool post and fed axially into the partially opened chuck jaws. It is important when doing this to lock the chuck jaws to prevent the scroll opening the jaws further while grinding is in progress. This is rather similar to the procedure described earlier for boring the soft jaws on the 3-jaw chuck. Since trueing the jaws in this way, the chuck performs extremely well and will centre round stock repeatably to within a small number of thou.

When performing any grinding operation like this in the lathe, it is important to protect all sliding surfaces from dust, as any abrasive particles that find their way into the slides or onto the bed will shorten the life of the lathe significantly.

A tailstock travel indicator

I have to thank Tom Benedict, a fellow Taig enthusiast from the USA, for this particular idea, and for the excellent photographs of how he did it.

The tailstock in its standard form carries no indication of the distance of travel of the tailstock ram; when performing drilling operations in the lathe, such an indication is a distinct advantage if you require holes of a known depth. The sophisticated (i.e., expensive) solution to this kind of problem is to fit digital readouts that make use of graduated scales. The simple and crude solution is to engrave witness marks on the ram at known intervals.

Tom's solution to this problem was to attach a relatively inexpensive, 2" travel, dial indicator to the tailstock, with its plunger in contact with a striking plate attached to the tailstock ram. The general arrangement can be seen in Figure 79.

The dial indicator is mounted on a block of aluminium (see left hand picture); two counterbored UNF10-32 screw holes 1" apart provide for T-bolts/nuts to attach the mounting block to the rear of the T-slots on top of the tailstock. A hole is drilled and reamed longitudinally through the aluminium block to take the dial indicator, and a UNF 10-32 set-screw holds the indicator in place. The tip of the D.I. rests against a piece of 1/8" thick aluminium with a hole bored in it to fit the brass collar on the tailstock ram (see right hand picture). The outline was cut with a jeweler's saw, and the whole thing was filed to shape. A UNF 2-56 screw acts as a pinch bolt to attach the striking plate to the brass collar on the end of the tailstock ram.

Figure 79 - Tailstock travel indicator

Tom has applied the same measurement technique to give position indication on his cross slide and saddle, and also to the vertical milling slide—see Figure 80. One of the great advantages of these indicators is that they indicate the actual position of the components and do not suffer from the backlash associated with the lathe's leadscrews. A similar thing could be done using the now readily available digital scales.

Figure 80 - Lathe fitted with position indicators

Chapter Nine
Dividing and graduating

One of the things that it is often very useful to be able to do with a lathe is to use it to rotate a workpiece by a known angle, or a known fraction of 360 degrees, so that machining can take place at regular divisions around the work. Examples of this are cutting gears, machining flats on round stock, and graduating handwheels. Graduating has been mentioned a couple of times in *Enhancing the lathe* on page 88 onwards.

There are essentially three kinds of approaches that can be taken in order to do this kind of work in a lathe:

— Add a "division plate", which is essentially a disc with holes in it, spaced regularly around a circle, to the lathe headstock, and provide a means of indexing those holes with a plunger or detent of some kind.

— Acquire or build a "dividing head" or a "rotary table", which is essentially a spindle with a worm drive attached to it; the worm drive, along with a graduated handwheel, or more division plates, allows the spindle to be rotated by a known angle or a known fraction of 360 degrees.

— Add a dividing attachment to the lathe headstock, effectively converting the headstock into a dividing head.

Having some kind of dividing capability will greatly enhance the versatility of the lathe, and with it, the range of projects that can be tackled.

This chapter looks at a variety of ways that the Taig lathe can be enhanced for division, and how its own headstock (or a Taig headstock bought as a spare) can be converted into a useful dividing head.

Simple dividing operations

The simplest way of dividing in the lathe is to make use of the characteristics of the lathe itself and its standard accessories. For example, the 3-jaw and 4-jaw chucks give immediate dividing possibilities based on the number of jaws; 2, 3, and 4 divisions are possible, simply by holding the part in the chosen chuck and locating each jaw (or alternate jaws to get 2 divisions) in turn at a fixed point. The easiest way to achieve this is to cut a small "prop" that is the right height to fit between one of the chuck jaws and the lathe bed when the jaw is horizontal - i.e., cut the prop to exactly centre height minus half a jaw thickness. Alternatively, cut the props from threaded rod, make them slightly undersize, and add nuts at both ends to provide adjustment.

Two of these props, one for each side, will very effectively lock a 4-jaw chuck in position, which makes any machining operations (for example, using the filing rest described on page 82) a little simpler. So, if you wanted to file a square on the end of a part, for example, using the filing rest described on page 82, or mark 4 graduations on the periphery of a knob, indexing using the 4-jaw chuck and a pair of jaw props is an obvious solution.

Figure 81 - Propping the chuck jaws

With the 3-jaw chuck, some means of holding the chuck in position will be needed. One approach here is to wind a string round the chuck and hang a weight from it (the 4-jaw chuck?) to lock the jaw against the prop. 6 divisions can also be made with the 3-jaw chuck; you simply alternate between propping the jaw at the front and at the rear of the lathe bed. However, for these divisions to be accurate, you must make sure that the prop heights are adjusted to ensure that the propped jaw is parallel to the lathe bed. Figure 81 illustrates these jaw props in action with the three and four jaw chucks.

So, 2, 3, 4, and 6 divisions can be achieved very simply just by using the properties of the standard lathe chucks. Getting 5, 7 and higher numbers of divisions requires more work.

The next approach is to attach something else to the lathe spindle that has regular divisions around its periphery, for example, a gear or a division plate, and use some kind of detent to locate the spindle at each division (each gear tooth or hole in the plate). There are various ways that this can be done; Figure 82 gives an example of a setup used (and sold) by Nick Carter, where the division plate is permanently attached to the back of the spindle pulley, between the pulley and the headstock, and a plunger assembly is attached to the top of the headstock that can be adjusted so the plunger locates in the holes of one of the three hole circles. Any number of divisions that is a factor of the number of holes in the chosen hole circle is then possible; for example, if there are 30 holes in the circle, then 2, 3, 5, 6, 10, 15, and 30 divisions are possible, by counting the appropriate number of holes between divisions.

Figure 82 - Headstock division plate

Clearly, a drawback of the division plate approach is that you first have to acquire an accurately machined division plate; however, these can be obtained from the usual machinery suppliers, and can even be made yourself by careful use of a milling machine's X-Y table, or a bench drill and cross vice, if you take the trouble to calculate the necessary drilling coordinates for each hole. This is very easy to do if you have access to a CNC mill, or access to an existing dividing head or rotary table.

Making a division plate can be somewhat simpler if you have a few gears to hand, for example lathe change wheels; hold the division plate blank in the lathe chuck, and attach the gear to the rear end of the spindle by means of an expanding collet (see *Handwheel adapter* on page 101 for an example of such a collet). The teeth of the gear can then be indexed by a suitable detent, and the holes drilled in the plate by using a mini-drill (such as a Dremel, for example) clamped to the lathe cross slide.

If nothing suitable is available that will give the required number of divisions, a division plate can be improvised by careful marking out on the periphery of a disc. One way to do this is to cut a strip of paper that is the same length as the circumference of the disc, mark the divisions on this, and then wrap the paper strip around the edge of the disc. The marks can then be transferred to the edge of the disc.

In order to ensure that the final machining is as accurate as possible, make the improvised division plate as large as you can; any errors in marking out are then divided in the ratio of the radiuses of the plate and the part being machined.

Graduating using the lathe

Once you have devised a means of creating the desired number of divisions, and locking the spindle in the required positions, it is straightforward to cut graduations in a workpiece, either in the face of the piece or around its periphery, by holding a suitable cutting tool in the lathe tool post and either traversing the saddle with the saddle handwheel (to mark the periphery of the piece) or traversing the cross slide (to mark the face). A scriber can be used as the cutting tool for this kind of work - as with all tools, it should be held in the tool post so that the cutting point is at centre height. Once the graduations have been cut, numbers can be punched next to the marks if so desired.

Dividing heads and rotary tables

There is a limit to what can be achieved with devices that use simple division plates (generally known as indexers), as the set of divisions that can be achieved is limited to the factors of the hole circle that you are using.

A much more versatile dividing capability can be achieved by using a worm drive in combination with a graduated handwheel or a division plate. The idea here is that the worm gear is attached to the shaft that you want to rotate to the desired position (the output shaft), and a second shaft (the input shaft), at right angles to the first one, carries the worm and the division plate or handwheel.

The worm drive itself will enable a number of divisions to be achieved simply by making complete revolutions of the input shaft; a 60-tooth worm wheel will offer 2, 3, 4, 5, 6, 10, 12, 15, 20, 30, and 60 divisions by this means. Half-turns of the worm can be achieved simply by attaching a crank handle to the input shaft and sighting against a straight edge; this extends the range of divisions of our 60-tooth worm drive to cover 8, 24, 40 and 120 divisions as well.

Add a graduated handwheel to the worm shaft with, say, 60 graduations marked on it, and each division now represents a 10th of a degree of movement of the output shaft. The combination of handwheel and worm now offers all divisions that are factors of 3600 (60 graduations, multiplied by the worm drive ratio of 60:1), 43 different factors in all.

Devices that use worm drives to perform division are usually known as *dividing heads* or *geared rotary tables*. The main difference between the two is that a dividing head is generally used with its output shaft horizontal, will generally carry a chuck of some kind, and will often have a matching tailstock to support long workpieces held in the chuck, whereas a geared rotary table has an output shaft that is vertical, and carries a large circular plate with T-slots, much like a lathe face plate, as the work holding device. Having said that, rotary tables can often be used with their output shafts horizontal, and often come with adaptors that will carry chucks.

Figure 83 shows a simple dividing head that I designed for use with the Taig lathe; the view is taken from the rear of the lathe for clarity. Instead of a graduated handwheel on the input shaft, this dividing head uses a fixed division plate with several circles of holes drilled in it, and an indexing arm attached to the input shaft, which carries a plunger detent that can engage with the holes in a chosen hole circle to lock the indexing arm in position. The combination of the worm ratio and the number of holes in the chosen hole circle gives further division possibilities; hence, with the right hole circle, the chosen number of divisions can be achieved.

Figure 83 - A dividing head

The output shaft of this dividing head can carry a Jacobs chuck, and is at the correct height such that when the dividing head is mounted on the Taig lathe cross slide, the chuck is at centre height. This makes it possible to use the dividing head to drill radial holes in a workpiece, with the axis of the output shaft at right angles to the axis of the lathe spindle. In the photo, the dividing head is mounted on the Taig milling slide and is being used to cut a clock gear; the output shaft holds the gear blank, and the lathe spindle holds a multi-tooth gear cutter. To cut a tooth, the dividing head is moved vertically down so that the blank is passed "through" the cutter, and then returned to its original position. The dividing head is then used to position the blank to the next division, and the process repeated.

In order to use a dividing head to generate a given number of divisions, you need a hole circle (or a graduated handwheel) with a number of holes in it that, multiplied by the worm ratio, gives a number that can be divided exactly by the desired number of divisions. For example, you have a hole circle of 45 holes, and a worm ratio of 30:1. Moving the indexing arm one hole in the hole circle causes the output shaft to move by 1/1350th of a revolution. If you desire to create 50 divisions, this is possible, as 1350 divided by 50 gives 27 exactly; hence, advancing the indexing arm 27 holes for each move generates 50 divisions. However, if you wanted 19 divisions, this combination of division plate and worm ratio is not suitable, as 1350 divided by 19 gives a remainder. The only solution would be to use a hole circle with a multiple of 19 holes in it; for example, with a 19-hole circle and a 30:1 drive ratio, moving 30 holes (1 complete revolution plus 11 holes) for each move would give 19 divisions.

Headstock dividing attachment

Figure 84 - Headstock dividing attachment

The use of a division plate to index the lathe spindle has been described earlier in this chapter. However, for some operations, it can be convenient to turn the lathe into a dividing head, by attaching a worm drive to the lathe spindle. This is particularly convenient when an indexing operation is needed on a part that has just been machined in the lathe, as the part can be left in the chuck undisturbed and will therefore be guaranteed to be on centre. Examples of this kind of use include:

— Machining gear wheels, where the gear blank has been turned to the right diameter in the lathe, and the gear cutting is then performed by means of a secondary spindle that carries the gear cutter.
— Graduating handwheels, as discussed earlier.
— Machining division plates.

Figure 84 shows a dividing attachment that I designed to be fitted to the Taig lathe headstock, and which will be described as a constructional project in the rest of this chapter. In the photo, it is attached to a "spare" Taig headstock that has had its "legs" shortened and a base plate added; this means it can be used as a stand-alone dividing head that will mount on the Taig cross slide with the output shaft at centre height. However, the dividing attachment can just as easily be fitted directly to the lathe headstock, as the attachment is secured via the T-slots in the top plate of the headstock.

A 30-tooth worm gear is attached to the rear end of the lathe spindle, in place of the lathe pulley, and the input shaft is held in place by means of a bracket attached with bolts to the headstock T-slots. A brake shoe is also present, in between the headstock and the worm wheel; this allows the spindle to be locked in position after a move. The brake shoe is held in place with a pair of cap head screws.

This dividing attachment design includes an additional "twist", as there is a second 30-tooth worm wheel attached to its division plate carrier, allowing the division plate to rotate under the control of a secondary input shaft, which carries a graduated thimble with 40 graduations on it. When the indexing arm of the dividing head is engaged with one of the holes in the division plate, turning the thimble causes the division plate to rotate, which in turn causes the indexing arm to rotate. This combination therefore gives 1/100th of a degree of movement of the output shaft if the graduated thimble is rotated by one division. The advantage of this "compound division" arrangement is that, for numbers of divisions that are not often used and for which there is no convenient division plate available, the angle between divisions can be calculated, and translated into whole turns of the indexing arm plus whole and part turns of the thimble.[12]

For example, suppose we wish to cut a gear with 47 teeth, and we don't happen to have a division plate that will give this number of divisions in conjunction with the 30:1 worm drive. 360 divided by 47 is 7.659, so, rounded up to the nearest 1/100th of a degree, a movement of approximately 7.66 degrees is needed between divisions. This will give an overall error of 2/100ths of a degree, so the last division will be 7.64 degrees instead of 7.66; we might choose to make the 23rd division 7.65 degrees in order to make the maximum error 1/100th of a degree for any one division. 7.66 degrees is less than a full turn of the indexing arm (12 degrees per turn), so each division requires 19 full turns of the thimble, plus 15 graduations (14 for the two "thin" divisions). Rather than have to turn the thimble 19 times per division, it is more convenient to fit a 30-hole division plate, and to rotate the indexing arm by 19 holes in the plate instead. The sector arms in front of the division plate are adjusted to this 19 hole separation, to avoid the operator having to count holes (and then get it wrong!). Similarly, if you are using the graduated thimble, it is wise to write down the sequence of actual readings that you expect to see on the thimble for each move, rather than calculating so many graduations on from the current position each time, unless you are very confident of your mental arithmetic, and more importantly, very confident of your ability to maintain concentration during a repetitive task.

---

[12]This is a variant of the design used by George Thomas in his celebrated "Versatile Dividing Head" design, described in his excellent book, "Dividing and Graduating", published by Tee Publishing (ISBN 0-905100-85-9). The VDH design uses two 60:1 worm drives, and 100 divisions on the graduated thimble, giving the ability to set the angular position to a thousandth of a degree. I felt that this was overkill and settled for a positioning capability of only 100th of a degree per division, which is as good as is needed for most purposes.

Note that when working out how many holes to move the indexing arm, you need to ignore the hole that the arm is currently located in. In other words, to adjust the sector arms to the right separation, they should uncover one more hole than the number of holes to be moved; the first uncovered hole is the starting position and the last uncovered hole is the target position for the move. Getting this wrong will result in a bad case of "the thin tooth problem" if you are cutting a gear, and needless to say, much frustration and wasted effort.

By this approach, it is possible to create any number of divisions that you care to name, and any one division will at most be 1/100th of a degree larger or smaller than it should be. This level of error is small enough to be ignored for most practical purposes.

Building the dividing attachment

As mentioned above, the drive ratio for the main worm drive is 30:1, using standard parts from HPC Gears (see Other suppliers on page 186). This choice of ratio was largely dictated by the physical dimensions that I was working within; I wanted the dividing attachment to be usable as a dividing head mounted on the lathe cross slide, with its spindle at centre height, and there is approximately 1¼" between the spindle and the surface of the cross slide. Limiting the height of the dividing components was also desirable given that this would make it suitable for use in the Taig mill as well, as it has relatively limited Z-axis travel. The final choice was a 30T, 20DP phosphor bronze worm wheel and single-start steel worm (HPC part numbers M20-30 and W20-1 respectively)[13]. This seemed to provide a drive that fitted the space available and was also substantial enough to take the 5/8" bore required to fit it to the Taig spindle. The choice of drive ratio was also affected by other considerations; with a 30:1 ratio, it is easy to cut a 12-hole circle, and from that, to cut a 360-degree protractor. Once you have a protractor fitted to a dividing head, it is then a simple matter to cut any of the other hole circles that you may decide are useful.

The design for the compound division components uses a supplementary 30:1 worm drive to position the division plate, and 40 divisions on the graduated thimble

---

[13] HPC also supply their worms and wheels in Delrin, at a reduced price and with somewhat lower torque capacity (2.75 Nm as opposed to 11.41). It may well be viable to make use of these parts in some applications, particularly if the spindle will always be locked prior to machining.   A halfway house offering an intermediate torque capacity of 7 Nm is to use a steel worm with the Delrin wheel.

fitted to the secondary worm shaft, giving positioning of the output shaft to a theoretical resolution of 1/100th of a degree as mentioned above. The ultimate positioning ability of a device built to this design will be limited by the accuracy of the individual components and of the overall construction, so whether positioning to this accuracy is really achievable is another matter altogether. As this secondary worm drive will never be required to take any significant load, a Delrin worm drive from HPC's 32DP range (HPC part numbers ZPM32-30 and ZW32-1 respectively) is used for this component.

Further features of this design include the ability to perform direct dividing by attaching division plates directly to the spindle in place of the worm drive. Although not described here, rising blocks could easily be added for machining diameters in excess of 2.5". Similarly, tailstock support can be added for situations where it is necessary to machine long parts.

Finally, wherever possible, the components have been designed so that the head can be assembled with either "handedness". When working on the lathe cross slide, it is often convenient to have the spindle nose pointing left with the division plate facing the operator (this is the configuration shown in Figure 84). However, when using the components with the lathe headstock, it would be more natural to have the division plate facing the operator with the spindle nose pointing right, as is normal for the lathe. Converting between the two orientations can be achieved very rapidly.

So, in summary, the design as described below offers a wide variety of interesting possibilities from which the reader can pick and mix to suit particular needs:

— "Direct" dividing, using a simple division plate and detent.
— "Simple" dividing, using a 30:1 worm drive and a division plate.
— "Compound" division, using two 30:1 worm drives in tandem.
— Use of the simple and compound division components with either "handedness".
— Using any of the above options either directly with the Taig lathe headstock itself, or with a separate spindle, comprising a second headstock, optionally modified for cross slide mounting with its spindle at centre height.
— Use of the dividing spindle as a cross slide or milling slide mounted milling spindle (with a pulley set and suitable drive motor attached).
— Use of any of the above with tailstock support. A by-product of the tailstock design described is that it will conveniently double as a boring bar holder for use on the Taig lathe's cross slide.

Of course, the usual comments apply at this point; the materials, dimensions, etc., shown in this design are in many cases simply a reflection of the materials the author had to hand, and can therefore be modified and substituted to suit the contents of your own stock and scrap box.

Modifying the headstock

If it is desired to use the dividing attachments and Taig headstock as a stand-alone dividing head, there are two options:
— Use an un-modified Taig headstock (maybe "borrowed" from your lathe, or a second one purchased for the purpose). If you plan to take this option, ignore the instructions below for modifying the headstock, and continue at *The brake shoe* on page 120. The un-modified headstock can be mounted by means of a short dovetail plate – effectively a short piece of lathe bed – as it is on the Taig Mill.
— Purchase a second Taig headstock and modify it to be usable on the Taig lathe cross slide at centre height, as described below.

The Taig headstock consists of an aluminium extrusion that holds a steel spindle carried in a pair of substantial 4cm OD ball races, with a steel spacer tube between the two races. The lower part of the extrusion forms two "legs" that carry dovetail slots; these locate the headstock onto the lathe bed. A 10-32 UNF cap screw tightens the clamping piece at the back of the headstock onto the dovetail slots of the lathe bed. The spindle and bearing assembly is press-fitted into this extrusion, so it should be left in place while the modifications are performed[14].

Taking your courage in both hands, remove the lower ends of the "legs" with a hacksaw, as shown in Figure 85. This will remove the dovetail slots, leaving a pair of "feet" a few millimeters long below the bearing housing.

---

[14] If you are modifying an old-style headstock, where the body is in two parts, the headstock can be disassembled for leg removal and re-assembled before milling the legs to length. The diagram in Figure 80 actually shows the old-style headstock, but the intent is the same, regardless of which is used.

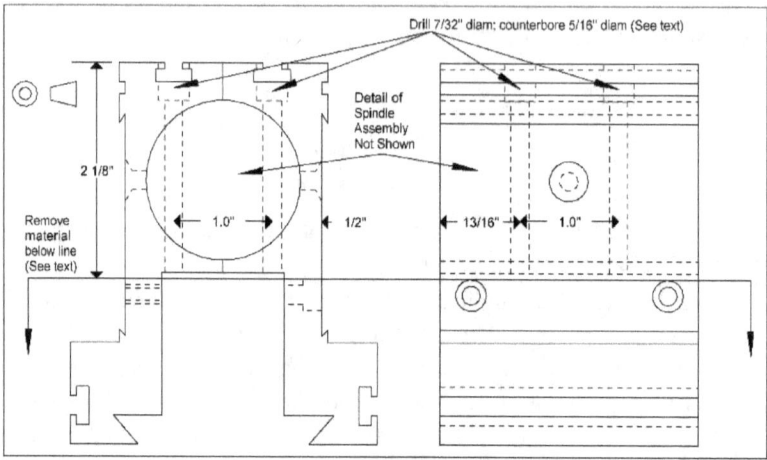

Figure 85 - Taig headstock modifications

The two "feet" can now be milled (or carefully filed) to their final height, which should leave about a millimeter of each foot below the housing. The final dimension after machining is far from critical at this point, as you get a second bite at the problem later on when the base plate (Figure 86) is added and machined to give the right final centre height. The requirement here is to create a nice flat base onto which the base plate can be attached. I found that the most convenient way to mount the headstock for this operation was to use the T-slots in the top surface of the headstock extrusions as a means of attaching a 3" by 2" mounting plate with countersunk T-bolts. The assembly can then be fixed to a milling table (or the Taig vertical slide) with further T-bolts through the mounting plate.

The base plate is made from a piece of 3/8" by 2" aluminium, 2 5/8" long; it is fixed to the newly milled feet on the bottom of the headstock by four 4BA cap head screws. Ideally, the holes for these screws and the threaded holes in the feet should be drilled together. Clamp the base plate in position on the base of the spindle assembly and drill through both items with a 3.1mm (4BA tapping) drill. The holes in the base plate can then be opened out to 3.6mm (4BA clearance), and counterbored 6mm to bury the cap heads well below the surface. Tap the holes in the feet, then fix the base plate in position with the cap head screws. Note that the hole positions are not symmetrical; this is simply to ensure that should the base plate be removed for any reason, it will always go back the same way around, although there is not likely to be very much reason to do this.

THE TAIG/PEATOL LATHE

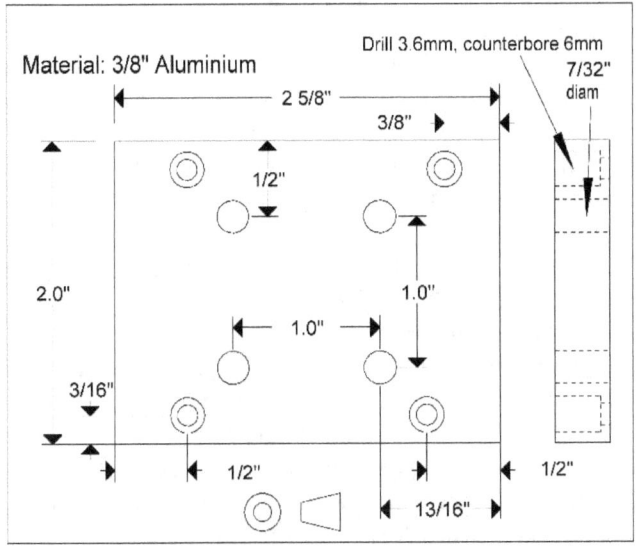

Figure 86 - Base plate

The remaining four holes in the base plate are to allow T-bolts to pass through the base; these holes are arranged in a 1" square to match the spacing of the T-slots on the Taig lathe and mill. By now you will have spotted that these T-bolts must pass right through the body of the truncated headstock. The easiest way to drill these holes is by mounting the spindle/base plate assembly upside down, as it was for milling the feet, and to drill through from the base using a 7/32" drill. Careful marking out is needed here; the objective is to have the T-bolt holes straddle the spindle, and to emerge in the middle of the T-slots in the upper surface of the headstock housing. In the process, the drill will partly pass through the steel spacer that separates the two headstock bearings, as well as the aluminium extrusions, but will not penetrate into the bore or damage the spindle itself. This drilling operation provides ample opportunity for the drill to deflect, so gentle drilling and good lubrication is the rule. The assembly is then reversed, and the four holes counterbored 5/16" diameter to a sufficient depth to allow a 10-32 UNF cap screw head to sit below the bottom of the upper T-slots.

The assembly can then be returned to the mill or vertical slide to skim the base plate; this can be seen in progress in Figure 87. The plan here is to fix the spindle height of the dividing head such that it will be on centre height when fixed to the cross slide. Nominally, centre height over the cross slide is 1 ¼". However, it is worth doing some careful measuring to check the height on your own lathe. It is also worth taking care in the set-up to ensure that the final machined base is actually parallel to the axis of the spindle, and square to the sides of the spindle housing.

The T-bolt holes needed to secure the spindle to the cross slide or milling table are rather long. You will need 2 1/4" 10-32 UNF T-bolts to fix the spindle to the Taig lathe cross slide or 2 ½" bolts to fix it to the Taig mill table (which has the same 1" T-slot spacing as the lathe, but the slots are deeper.) If bolts of this length are hard to locate (as is the case in the UK/Europe), then 2BA is a close substitute, or alternatively 5mm.

The result thus far is the first usable product of this project – a milling spindle that can be mounted on the lathe cross slide or vertical slide (or indeed, on a milling table if you can dream up an application for this!). The Taig lathe pulley sets and drive belts can be pressed into service for this purpose, along with a suitable fractional horsepower motor. I have not attempted to describe mounting arrangements for this application, as motor mountings and dimensions will vary. However, there is plenty of scope, given the availability of the two T-slots on the top of the spindle housing, to provide a mounting that can allow a motor to be rapidly attached and detached as needed. The short drive belts supplied for the Taig Mill are probably the most appropriate for this application; using a Taig pulley set, this combination gives approximately 3.8" between the axes of the milling spindle and the motor shaft.

Figure 87 - Skimming the base plate

The brake shoe

It was difficult to decide on the best approach for providing a locking mechanism for the spindle. Clearly, if one was building the spindle and housing from scratch, the obvious approach would be to fit some kind of clamp between the two bearings. However, as this would mean dismantling the pre-loaded spindle assembly in order to modify its steel spacer, this was not an easy option in this case. Remaining options were to fit a brake shoe at the nose (turning down the 1" AF hex nut behind the nose thread to form a brake drum), or at the rear of the spindle, directly behind the rear bearing and pre-load nut. The latter approach was adopted, as it is extremely useful to be able to use a spanner on the spindle when tightening the collet closer.

The rear end of the spindle poses problems of a different kind. There is a circlip that ensures that the spindle cannot escape the clutches of the bearings; this protrudes slightly less than 2mm proud of the aluminium housing. The rear bearing itself protrudes somewhat less than this also. So, the brake shoe has to be offset from the rear surface of the housing by 2mm. The spindle itself is 17mm diameter at the point where the circlip retains it in the bearing; it then rapidly changes to 5/8" diameter for the remainder of its length. The brake shoe needs to be bored with both 17mm and 5/8" diameters, so that it can straddle the change in diameters of the spindle. This allows the available space at the rear of the spindle to be used to its best advantage, and in particular, also leaves sufficient room for the worm drive to be attached. These factors explain the rather curious shape of the brake shoe, shown in Figure 88.

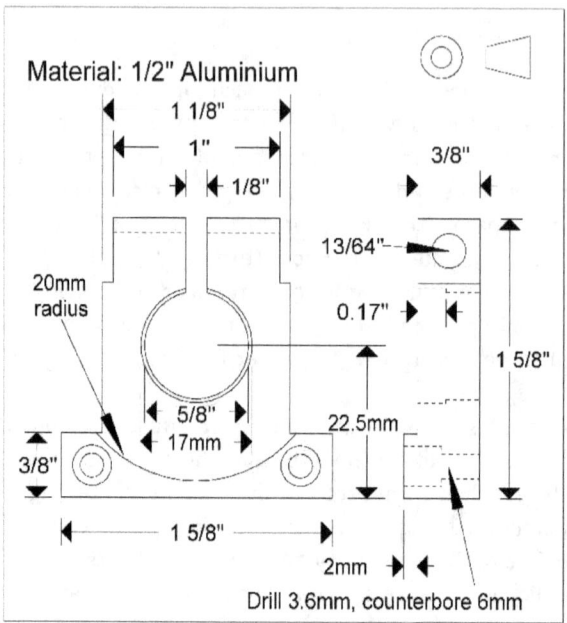

Figure 88 - Brake shoe

The shoe is made from a piece of ½" thick aluminium, 1 5/8" square. I had the luxury of being able to machine the shoe entirely using milling and drilling operations in my Taig CNC mill. Boring 17mm diameter holes with a 5/32" end mill is a lot of fun with such a machine; this operation can be seen in progress in Figure 89. Needless to say, the lathe, vertical slide and 4-jaw chuck are equally appropriate tools for the operations required here.

Figure 89 - Machining the brake shoe

The final shoe will be 3/8" thick, with an extra 2mm for the foot to offset it from the pre-load nut and rear bearing. Having trued up the piece to 1 5/8" square, it is necessary to mill or turn one face to reduce the overall thickness to 0.454" (3/8" plus 2mm). The foot is then formed by removing a further 2mm of material on a 20mm radius centred 22.5mm from one edge, and 13/16" from the adjacent edges. The piece is then through bored 5/8" diameter, on the same centre, followed by boring 17mm diameter to 0.17" below the surface. The profile of the piece is then cut as shown in Figure 3, and a hacksaw or slitting saw used to cut through to form the split clamp. Drill the mounting holes as shown, and counterbore them to allow the 4BA cap head mounting screws to be buried below the surface. Finally, drill the 3/16" hole for the pinch bolt. Note that the 3/8" square cutouts allow the pinch bolt (Figure 90) to be inserted in either direction through the clamping hole; this is one of the "ambidextrous" design features mentioned earlier. The cutout serves to prevent the 3/8" square pinch nut from turning – use a standard Taig lathe square T-nut for this purpose or make one up from 1/8" X 3/8" flat stock.

The pinch bolt and tommy bar

The pinch bolt (top of Figure 90) is made from a 2.25" length of 5/16" diameter mild steel. Turn down 1.25" of its length to 0.19" diameter and thread the last 3/8" 10-32 UNF. Cross-drill ¼" from the other end 3.2mm. The tommy bar (bottom of Figure 90) is a 1.5" length of 1/8" diameter silver steel rod ("drill rod" in the US); thread each end with a 4BA die for about 1/8". Strictly speaking, 1/8" is too small for 4BA threading, but it will take enough of a thread for this purpose. Fit a 4BA brass nut to each end and turn them down to remove the flats. Pass the tommy bar through the hole in the pinch bolt and fix the end caps in place with a drop of Superglue.

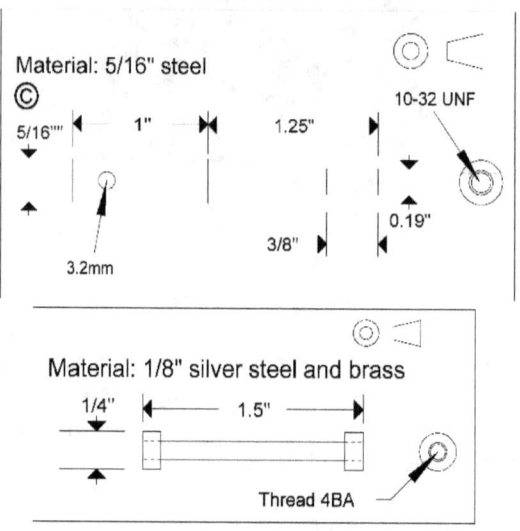

Figure 90 - Pinch bolt and tommy bar

The pinch bolt can then be fitted through the hole in the brake shoe, with a steel washer under the shoulder and a square 10-32 UNF nut (the type used as T-nuts on the Taig lathe) at the threaded end. As mentioned above, the bolt can be fitted from either direction, depending upon the desired "handedness" of the indexing components.

The shoe can now be fitted in place on the rear surface of the spindle housing, with its foot and mounting screws at the bottom.

Clamp the shoe onto the shaft using the pinch bolt and nut. Spot through the mounting holes, drill 3.1mm (4BA tapping), and tap 4BA. The shoe can now be screwed into position. Note that if you plan to use the dividing components on the lathe headstock, this operation will need to be repeated on the rear of the lathe headstock. Figure 91 shows the completed brake shoe assembly, and the 30T worm wheel, fixed in position on the rear of the spindle assembly.

Figure 91 - Brake shoe and worm wheel in place

The dividing assembly mounting plate

This component is shown in Figure 92; it serves as a universal mount for the dividing attachments. This plate is designed to fit at the tail end of the headstock assembly, using the T-slots in the top of the aluminium extrusion – hence the two 3/16" holes in the base plate to take 10-32 UNF T-bolts. The two semi-circular notches reduce the degree to which the plate obstructs the rear pair of T-bolts that pass through the headstock. If preferred, these could be elongated to merge with the 3/16" holes to form a slot; the only disadvantage with this alternative is that the T-bolts for the mounting plate are then no longer "captive".

The mounting plate is fabricated from 2" lengths of ½" X ¼" and 1" X 3/8" steel bar; these should preferably be clamped together while milled or filed to length in order to ensure that the lengths are exactly the same. However, this is only a cosmetic issue, as the length is not terribly critical. The lower 3 holes in the 3/8" thick vertical plate need to line up with the 4BA tapped holes in the ¼" thick horizontal plate. The intent here is that the two plates are clamped together using 4BA socket head screws, and the counterboring allows the heads of the bolts to end up below the surface of the plate, leaving the rear surface of the plate unobstructed. The usual approach of careful marking out, spotting through holes etc., should be sufficient to ensure that they line up. Those of you that have a mill available (or even a cross vice with graduated leadscrews) will find this straightforward as the hole positions can be dialed in from a reference edge.

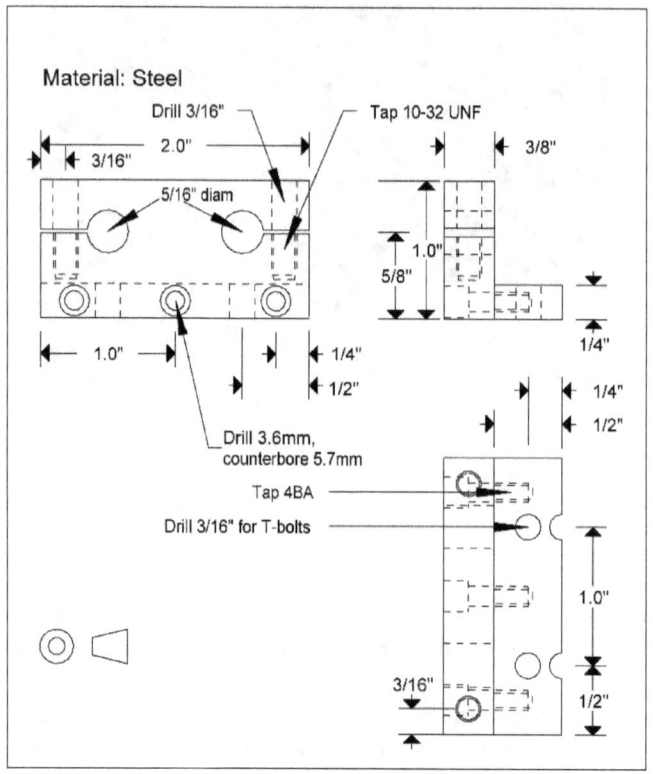

Figure 92 - Mounting plate

The two 5/16" holes should be drilled undersize and reamed to size; these need to be a close fit with the 5/16" steel bars that will be used to hold the various dividing components. The vertical holes for the 10-32 UNF pinch bolts are drilled last, after the horizontal saw slits have been cut. Drill to depth with a 10-32 UNF tapping drill (4.1mm) and then follow up with a 3/16" drill to the level of the saw cut. The lower holes can then be tapped UNF 10-32. If you feel particularly energetic, the pinch bolts can be replaced by variants on the pinch bolt and tommy bar shown in Figure 90; however, as these bolts are likely to be used relatively infrequently, this may not be felt worth the additional effort involved.

Modifying the 20dp worm wheel

Figure 93 - Worm wheel modification

The 20DP worm wheel as supplied by HPC Gears (see above) comes with a 3/8" diameter bore. This needs to be increased to 5/8" diameter to fit the end of the spindle. HPC will modify bore diameters for an additional fee, but this can also be achieved in the lathe. Grip the 1" diameter boss in the 4-jaw and centre it carefully using a dial indicator. Alternatively, the boss can be held in the 3-jaw, if it is sufficiently accurate, or if you are prepared to bore out the Taig chuck's soft jaws to hold the boss spot on. The bore can then be opened out using a suitable boring bar; this can be seen in progress in Figure 93, using a 1/4" boring bar fitted with a Sumitomo titanium carbide insert.

The wheel will need a set screw; drill and tap the boss 10-32 UNF to take a ¼" long hex socket head set screw. The wheel is fitted "boss first" onto the spindle and located with the grub screw. The wheel will overhang the end of the spindle by about ¼"; leave a small gap between the boss and the brake shoe so that there is no rubbing between these components. I chose to file a flat on the spindle to act as a pad for the set screw to locate against; I have filed a similar flat on my lathe spindle. This simplifies fitting and removal of the wheel and drive pulley, as the inevitable burrs raised by the grub screw do not then foul their bores.

The division components

The construction of the various support components for the two worm drives will now be described, followed by the division plates, indexing arm and sector arms. For those that are only interested in building the simple indexing capability (one worm drive), the construction can be simplified, and some components omitted; this will be detailed as the description progresses.

The primary worm carrier

Error! Reference source not found. shows this component. The carrier has two split clamps that will take 5/16" diameter rods; the left-most of these allows the carrier to be mounted in position, using a 1 ¾" long steel rod held in one of the corresponding clamps on the Dividing Assembly Mounting Plate (Figure 94). The right-most clamp in the figure can be omitted if only one worm drive will be used, as this is used to mount the secondary worm and its carrier.

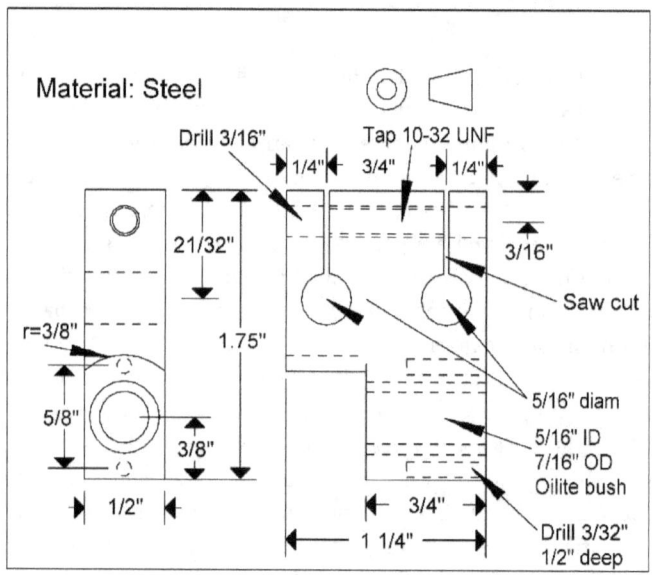

Figure 94 - Primary worm carrier

The carrier is fitted with an Oilite bush that forms a bearing for the primary worm shaft. Above and below this bush are marked two 3/32" diameter holes. These carry a pair of locating pins for the Index Plate Boss (Figure 101) which are needed when performing simple indexing without the secondary worm drive fitted; these holes can be omitted in both components if it is intended to leave the secondary worm drive permanently installed. The 3/8" radius shown above the bush gives clearance for the worm and its set screw and can be adjusted to suit. The hole for the worm shaft bearing is best drilled/bored in the 4-jaw chuck on the lathe to ensure that it is perpendicular to the face of the carrier; the 3/8" radius cutout can be machined at the same setting. Construction of this component is otherwise straightforward.

The secondary worm assembly

The Secondary Worm Carrier (Figure 95) and the Worm Lock (Figure 96) are machined in the following sequence.

First, cut a length of 3/8" diameter brass rod to 23/32" long (1/32 less than ¾"), drill it axially and tap UNF 10-32 for its entire length. Next, cut a piece of ¾" square section steel bar to length (1 ¼") and carefully mark out the positions of the four holes. The bar can now be mounted in the 4-jaw chuck, and accurately centred to bore the 3/8" diameter hole for the worm lock. Remove the work from the chuck – loosen two of the four jaws, to make the next re-positioning operation easier. Now temporarily fix the brass rod in the hole with Superglue, taking care to position it so that it is equidistant from each face of the steel bar. If you avoid de-greasing these components too thoroughly before gluing, the Superglue will hold them together just fine for the next drilling operation but will allow the worm lock to be driven out afterwards with a suitable drift.

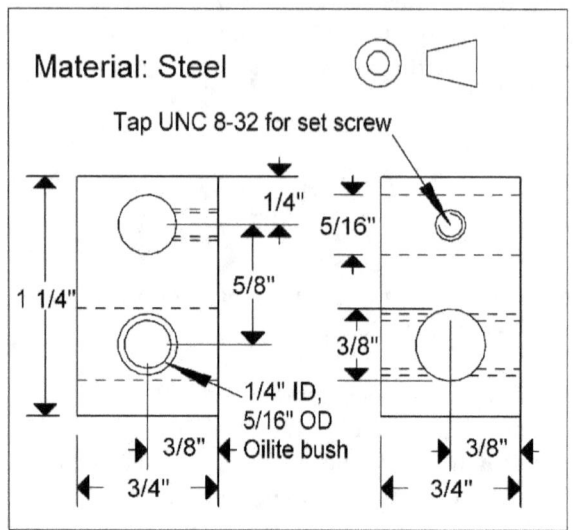

Figure 95 - Secondary worm carrier

Re-mount the piece to drill the hole for the Oilite bush; again, carefully centre the hole position. The hole is first drilled ¼" in diameter; this ensures that the hole in the worm lock will be concentric with the secondary worm shaft, regardless of marking out or machining inaccuracies. Remove the piece from the chuck and use a drift to separate the worm lock from the carrier. Clean off any glue residue. The ¼" hole can now be bored out to 5/16" to take an Oilite bush, and the 3/8" hole cleaned out with a drill and reamer to remove the central portion of the bush.

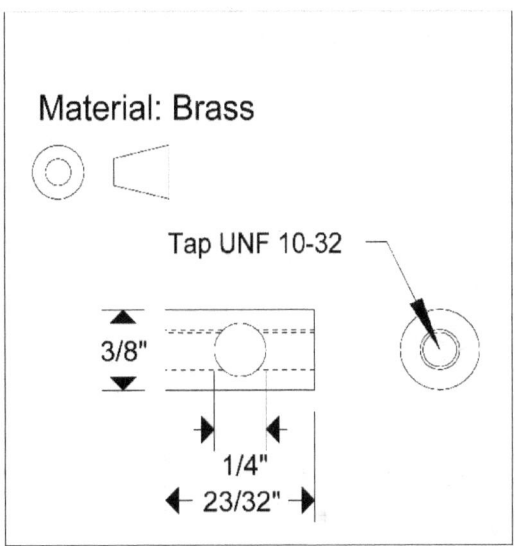

Figure 96 - Worm lock

The remaining 5/16" hole and the UNC 8-32 hole for the set screw complete the secondary worm carrier. A 1 ¾" length of 5/16" steel rod is used to connect it to the primary worm carrier; file a flat for the set screw 3/8" from one end. An alternative to the set screw is to fix the rod permanently using Locktite and a steel pin, as there should be no need to remove the rod once fitted.

The worm lock can be re-fitted to its hole in the carrier. If all is well, it should be possible to pass a length of ¼" diameter silver steel through the Oilite bush and the worm lock; this will form the shaft for the secondary worm.

A 2 ¾" length of silver steel rod is used for the secondary worm shaft; I chose to use Locktite in addition to a set screw to locate the worm on the shaft. The thumbscrew (Figure 97) provides the means of tightening the worm lock; this is straightforward to machine from ½" diameter steel. The thread can be formed by reducing the diameter of the round stock; alternatively, cut a 5/16" length of stock, axially drill/tap 10-32 UNF and Locktite a length of 10-32 studding in place, finishing the threaded portion to 5/32". The latter approach has the advantage that the threads run right up to the shoulder, which makes life simpler.

The thumbscrew is finished with a light knurl. Having threaded holes at either end of the worm lock allows the secondary worm to be mounted with either "handedness" on the primary worm carrier. The lock operates very effectively, requiring little finger pressure to lock and unlock the secondary worm. As there is little force on this worm drive in operation, all that is required is to prevent inadvertent re-positioning of the Worm Thimble (Figure 98).

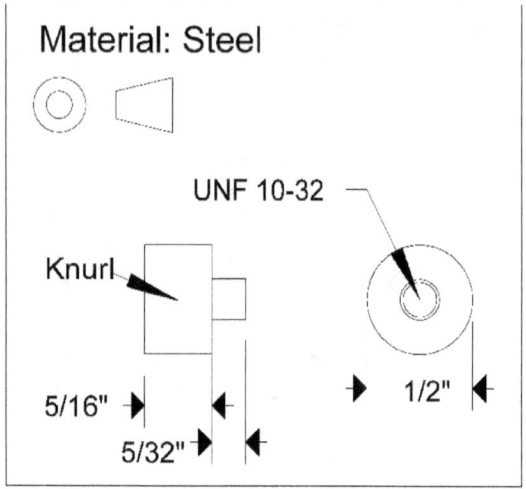

Figure 97 - Thumb screw

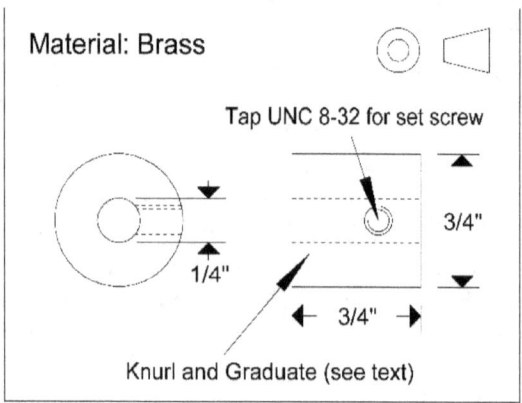

Figure 98 - Worm thimble

The Worm Thimble is machined from a piece of brass round stock, bored ¼" diameter and cross-drilled for a UNC 8-32 set screw. The surface of the thimble should be knurled at one end and marked with 40 graduations at the other end.

The final operation to perform on the secondary worm carrier is to add witness marks to aid with reading the position of the Worm Thimble. I chose to add these marks to the three possible faces; hence, the thimble position can be read from above the dividing head, or from the side, regardless of the "handedness" with which the components are assembled. I finished the steel components by using a chemical blackening kit; the black surface provides a useful contrast with the witness mark cut into the steel.

Figure 99 – Completed worm thimble

The thimble can be graduated by pressing the partially completed dividing head into service, once a division plate with a circle of holes that is a multiple of 4 has been made (as described below).

Having engraved the thimble with its graduations and numbers, it is worth making these marks stand out clearly against the metal background. A simple way of achieving this is to fill in the marks with indelible black felt-tip marker ink or black enamel paint, allow it to dry, and then carefully remove the excess from the high points with very fine emery paper, while the thimble is rotated in the lathe. The finished thimble can be seen in Figure 99.

The primary worm shaft

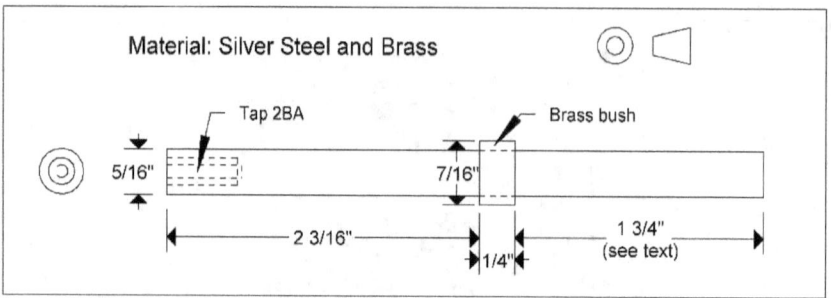

Figure 100 - Primary worm shaft

The Primary Worm Shaft is shown in Figure 100. This was cut from silver steel rod; the brass bush shown is Locktited in position. The end with the 2BA-threaded hole is passed through the Oilite bush in the primary worm carrier, from the right-hand side as shown in Figure 11, and the worm is fitted in position. The 2BA hole allows a screw and washer to be inserted in the end of the shaft as an aid to adjusting end-float in the shaft. Once adjusted, the worm's set screw can be tightened firmly – preferably against a suitable flat filed on the shaft to ease subsequent disassembly if need be. With the primary worm in position on its carrier, the dividing head can be partially assembled and can now be used to perform some of the later operations. However, the primary worm shaft should be left over-length at this point; its final length will be determined once the remaining indexing components are complete.

Index plate boss

The Index Plate Boss (which also carries the Secondary Worm wheel) is shown in Figure 101. This is machined from 1" diameter round stock; a suitable sequence is as follows. Cut and face a length of 1" diameter stock, sufficient to hold in the 3-jaw chuck with about 1 ¼" of stock overhanging. Reduce the diameter to ¾" over the last ¾" of its length. Bore the axial 5/32" hole, ensuring that the hole is a little over 1" in depth. Counterbore 7/16" diameter to a depth of just over ¼"; this should be a running fit with the bush on the primary worm shaft.

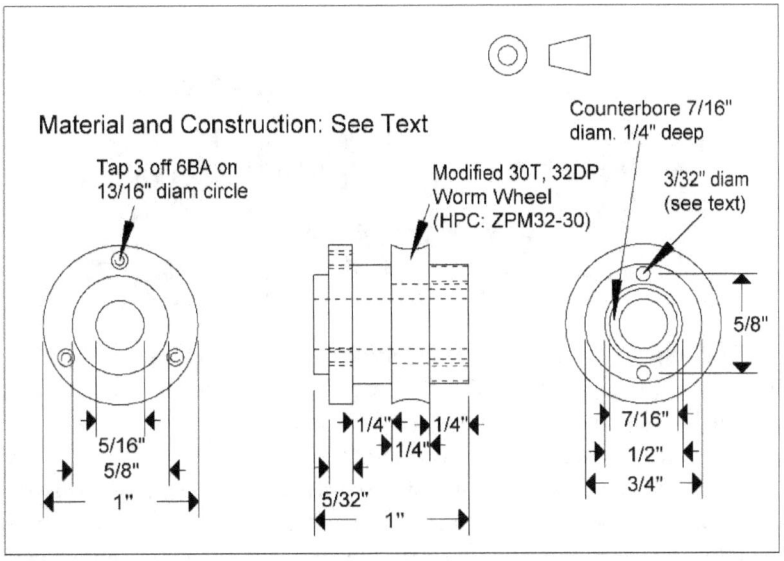

Figure 101 - Index plate boss

If you are implementing the secondary worm drive, further reduce the diameter to ½" over the last ½" of its length. This forms the shoulder that the secondary worm locates against. Part off to exactly 1" long. Bore out the axial hole of the 32DP, 30T worm wheel to ½" diameter, so that it is a push fit onto the boss. This operation is tricky - the specified worm wheel is made from Delrin, but with careful packing and centering in the 4-jaw, the wheel can be modified without damage to the teeth. Machine a ¾" OD, ½"ID bush ¼" long; I used brass for this, but the material used here is not critical. Assemble the modified worm wheel and then the bush onto the boss with Locktite, clamping them in position to make sure all is square while the Locktite sets. If the secondary worm drive is not being used, or if it is anticipated that the head will sometimes be used without the secondary worm fitted, the two 3/32" holes can be drilled in the other end of the boss. This can be done using the technique described below for drilling the three holes in the division plate flange, before parting off. A couple of ¾" long 3/32" diameter dowels can then be used to locate the boss in the corresponding holes in the primary worm carrier, preventing the boss from rotating. Of course, you must remember to remove these dowels before using the secondary worm.

Figure 102 - Drilling the flange for the division plates

The division plate flange is machined next. This can be achieved by using a short length of 5/32" diameter bar as a stub arbor, Superglued temporarily in place, allowing the boss to be centred in the 4-jaw (or in the 3-jaw if you have an accurate one). Having cut the flange, you can now demonstrate the usefulness of having a dividing head that can use the same accessories as the Taig lathe. Unscrew the chuck, still with the boss in position, and mount it on the nose of the embryonic dividing head. Mount the dividing head on the cross slide and position it to drill the three 6BA screw holes, using the Jacobs chuck mounted in the lathe headstock. The worm drive can be operated either with a temporary handle clamped onto the worm shaft, or by means of a screwdriver in the slot of the 2BA screw used for adjusting the end float. Figure 102 shows a prototype boss being drilled using this set-up. Accuracy in positioning these holes is important, as it is necessary for the holes to align up with the corresponding holes in the division plates, and for the division plates to be concentric with the boss when they are fitted in position.

The sector arms

Figure 103 shows the sector arm components, which can either be machined from 1/8" thick brass plate or from 1/16" plate components soft soldered together. The construction should be obvious from the drawings; the 1" OD, 0.8" ID portion of the right-hand arm is thinned to 1/16" and fits into the corresponding groove in the left-hand arm. The two 8BA holes are tapped through the thickness of the left-hand arm and are placed as close as possible to the 0.8" diameter. Two cheese-headed 8BA screws and washers clamp the two components relative to each other, allowing a chosen number of holes in a division plate to be exposed. It may be necessary to thin the left-hand arm underneath these two screw heads, or to apply a slight bend to the washers, in order to achieve the right clamping action. The arms can be used in reverse if necessary for large hole counts. The component sizes are suitable for the chosen division plate diameter of 3"; the dividing head will accommodate larger plates if need be, in which case the sector arms should be lengthened accordingly.

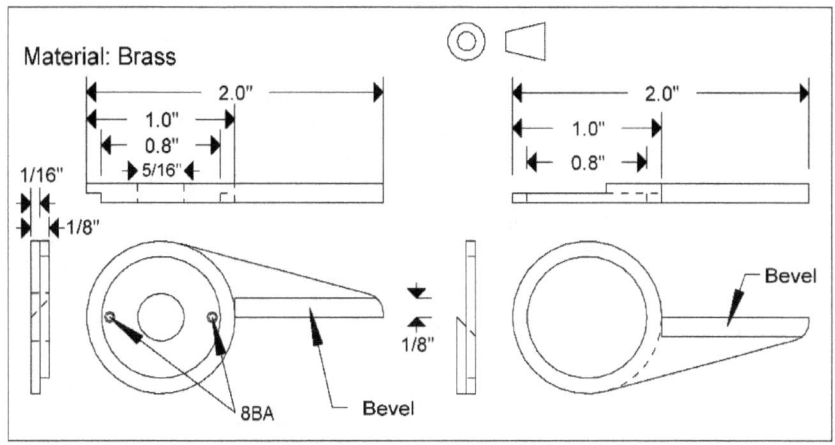

Figure 103 - Sector arms

The sector arms, and also the index plate boss, are retained in position by means of a small set screw collar, shown in Figure 104. Making this is a simple turning/boring job; the set screw used is 4BA and 1/8" long, seating in a flat filed in the primary worm shaft. This flat should be deep enough so that the set screw is flush with the surface of the collar when tightened in position. A split spring washer should be placed between the set screw collar and the sector arms and compressed before tightening the set screw; this will provide the necessary friction to ensure that the sector arms remain in the desired position while the indexing arm is rotated.

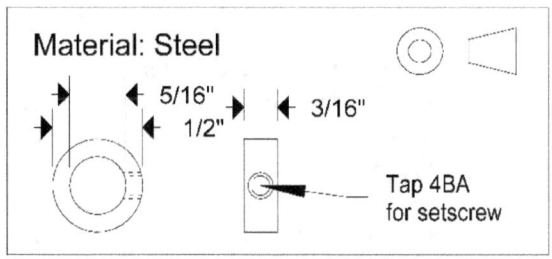

Figure 104 - Set screw collar

The construction could have been simplified by combining the set screw collar with the indexing arm clamp; however, the arrangement described allows the indexing arm to be removed and used for direct indexing operations without disturbing the rest of the dividing assembly.

Division plates and hole circles

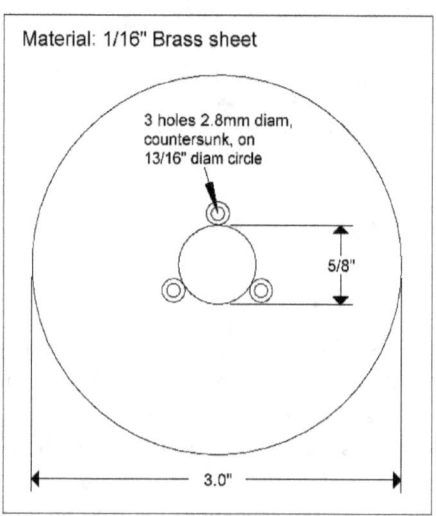

Figure 105 - Division plate blank

Figure 105 shows the design of the division plates. These are nominally 3" in diameter, as that was the size of a set of brass blanks I had to hand; the dividing head will accommodate larger plates if desired. The central hole size of 5/8" was chosen in order to allow the division plates to fit over the tail of the dividing head/lathe headstock spindle to allow their use for direct division, given a suitable mounting arrangement. The options for this kind of use are either to drill and tap the worm wheel to allow it to be used as a division plate boss, or to make up a 1" OD, 5/8" ID set screw collar drilled and tapped for the 6BA mounting screws.

A 3" diameter plate is not very useful for direct division when the dividing head is being used on the cross slide or a milling table without riser blocks, so it may be desirable to cut some smaller diameter plates for this purpose. However, for direct dividing with the lathe headstock, the 3" plates can be useful. It is also worth bearing in mind that the worm wheel itself has useful "real estate" available on it for a couple of low-numbered hole circles; I cut a 30-hole and a 24-hole circle in mine, as can be seen in progress in Figure 106. A further option for direct dividing in the lathe is to permanently attach a division plate to the inboard face of the lathe's drive pulley, as mentioned earlier (*Simple dividing operations* on page 108).

The decision as to how many plates to cut, and what numbers of holes to drill in each hole circle, will depend greatly on the applications to which the dividing components will be put. In order to make sensible use of the compound division capability, it is necessary to cut a 30-hole circle; a full rotation of the secondary worm is then equivalent to moving the indexing arm by one hole position.

Figure 106 - Drilling a hole circle in the worm wheel

Also, to mark the worm thimble with its 40 graduations, a 4-hole circle (or any multiple of 4) is needed. Arguably, as compound division will then allow positioning with a theoretical resolution of 1/100th of a degree, it may be that you will never need any further hole circles. However, as compound division involves keeping track of two independent movements, it is likely that users will find it convenient to have a set of hole circles that cover the most-used divisions for their applications. As cutting hole circles is very tedious, the best approach here is to cut the "obvious" circles first; those that allow generation of your most-used divisions. Leave the others until the need arises. A further possibility that can add to the versatility of this device is to mark one division plate as a protractor, with marks at one-degree intervals, and longer marks to denote the five and ten-degree points. This can be thought of as a simpler alternative to using compound division; in effect, it creates a 360-hole circle, giving 1/30th of a degree of output movement per degree of indexing arm movement. To create such a protractor, a 12-hole (or multiple of 12) circle is needed; this can be easily cut using the 30:1 worm drive, as 2.5 turns of the worm moves the spindle by 1/12th of a revolution. Moving the indexing arm by one hole in the 12-hole circle will rotate the output shaft by 1 degree. It is then straightforward to mount a blank division plate on the business end of the dividing head, mount the dividing head on the cross slide, and use a sharp scriber mounted in the lathe headstock to mark out the protractor.

Working out whether a given circle of holes can generate the requisite number of divisions is straightforward. Multiply the number of holes in the circle by the worm drive ratio; if you can divide the result by the desired number of divisions, with no remainder, then the chosen circle can be used successfully. For example, the four-hole circle will allow the dividing head to generate 40 divisions, as 4 X 30 gives 120, and 120/40 is 3 with no remainder. This also tells you that the indexing arm moves 3 holes between each division. The same principle applies when using direct division, except the drive ratio is 1:1 instead of 30:1.

The hole circles are easily cut using a 3/32" diameter drill – the most convenient approach here is to use a No.3 centre drill, which has this diameter of pilot drill. Using this size of hole, it is possible to fit an 80-hole circle onto the periphery of a 3" division plate. Leaving about 1/8" between circles, with care it is possible to squeeze up to 8 circles onto each plate; the smallest circle will be limited to a maximum of around 30 holes. It is worth remembering to mark the plate with the number of holes in each circle; counting holes several months later when you have forgotten what you did is a very tedious experience!

As mentioned above, compound division requires a minimum of two circles, one with 30 holes and one with a multiple of 4 holes; the latter in order to be able to mark out the graduations on the thimble. If you plan to cut a protractor, a 12-hole circle would usefully pass as a substitute for the 4-hole circle. It is useful to make up a mounting arbor to help with division plate drilling; this needs to have the right size of shoulder for the plates to mount on, the remainder being machined to suit your preferred method of work holding. The simplest approach is to machine a division plate arbor from one of the blank arbors that can be obtained for the Taig lathe (see "Blank and special purpose arbors" on page 26). The #1130 arbors are basically a short length of 1" diameter steel stock with a female thread at one end that fits the ¾" nose thread of the lathe spindle, and a 1" diameter washer held in place with a UNC 8-32 cap screw at the other. A 5/8" diameter shoulder can be machined on the blank end of the arbor; the washer and cap screw will then hold the plate firm for drilling. The arbor is then transferred to the dividing head for the plate drilling operations. Once the two initial hole circles have been cut in the first plate, the compound indexing components (and/or the protractor) can be completed, and then the dividing head can be used to cut any other size of hole circle that may be needed.

The indexing arm components

The left-hand diagram in Figure 107 shows the Indexing Arm Boss – a ¾" length of ½" diameter steel, bored 5/16" at one end and tapped 10-32 UNF at the other, and fitted with an 8-32 UNC set screw. A ½" diameter washer is also needed, shown in the right-hand diagram in Figure 107. The boss fits onto the end of the primary worm shaft, located by the 8-32 set screw. A 10-32 cap screw is used, with the washer, to clamp the indexing arm in position.

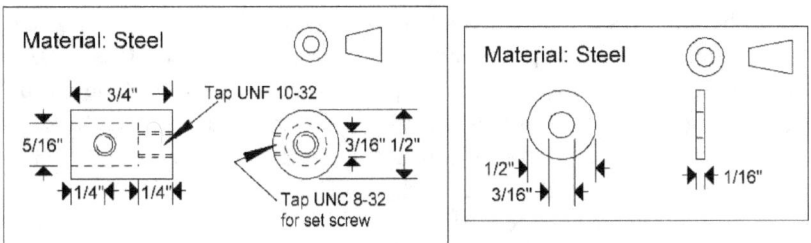

Figure 107 - Indexing arm boss and washer

The indexing arm itself is shown in Figure 107, cut from 1/8" thick brass plate. The 3/8" diameter hole takes the detent assembly; the longitudinal slot allows adjustment of the detent position relative to the division plate. The body of the detent assembly is shown at the top left of Figure 109. This is soft soldered into the 3/8" diameter hole in the indexing arm, ensuring that it ends up perpendicular to the arm.

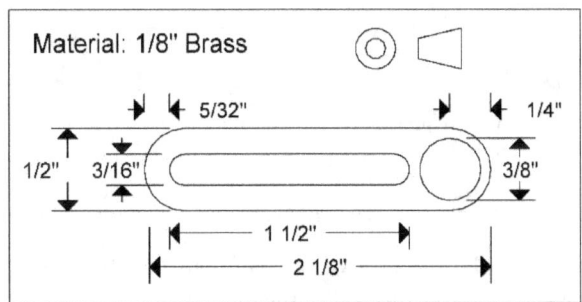

Figure 108 - Indexing arm

The detent plug (top right of Figure 109) and the detent knob (bottom left of Figure 109) are designed to allow the detent pin (bottom right of Figure 109) to be locked in the raised position. When moving the detent from one division plate hole to the next, you pull back the knob and rotate it half a turn to lock; another half turn allows the pin to be dropped into the next hole. A short spring will be needed to complete this assembly; it needs to be about ½" long, with an ID greater than 1/8" and an OD less than ¼". In operation, this spring will be required to compress by 5/16". The one I used happened to be on hand in my junk box; a suitable spring could be wound from piano wire if necessary.

The detent pin was made from a length of 1/8" silver steel rod ("Drill rod" in the USA). The last 1/8" is reduced to 3/32" diameter to match the size of holes drilled in the division plates; the diagram shows this as parallel turned, but a taper could be turned if preferred. The brass collar is soft soldered (or Locktited) in place ¾" from the tip of the pin. This component should not be cut to final length until the other detent components are complete and ready to assemble. A trial assembly will indicate the exact pin length required - fit the pin into the body, followed by the spring, the plug and the knob, the latter fitted in the lowered position. Mark the pin for cutting so that its end will be flush with the top of the knob. Final assembly can then be achieved by using a drop of Locktite to fix the plug into the body and a second drop to attach the knob to the pin.

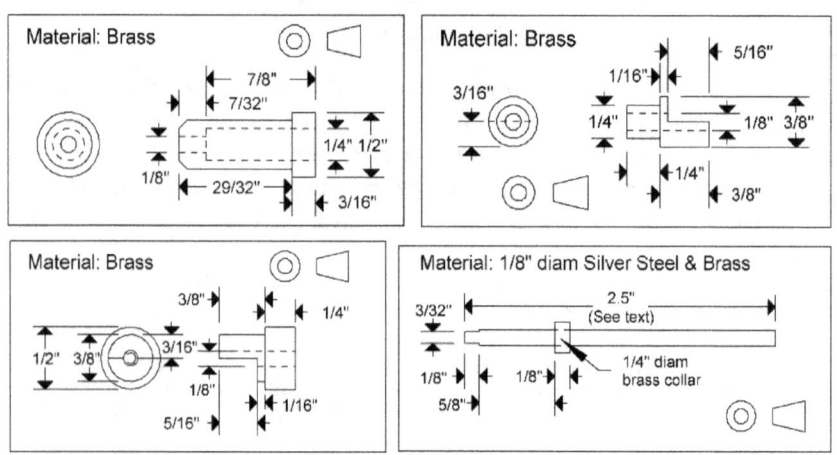

Figure 109 - Detent body, detent plug, detent knob, and detent pin

The final length of the primary worm shaft can now be determined. With the indexing arm boss in place on the end of the over-length shaft, attach the indexing arm and detent assembly and measure the distance between the shoulder on the point of the detent pin and the surface of the division plate. This tells you by how much the shaft should be shortened. Once the shaft is cut to length, file a flat to act as a pad for the set screw. I found that filing the last ¾" of the shaft flat formed a suitable pad both for this screw and for the set screw collar.

Tailstock support

For some operations, particularly when working on long items, it will be necessary to support the free end of the work piece using a tailstock of some kind. Clearly, if the dividing components are being used with an un-modified Taig headstock, it may be sufficient to press the Taig lathe tailstock into service for this purpose.

However, when using the headstock in its modified form, as described above, it may be necessary to make a tailstock that matches its reduced spindle height. The boring bar holder design (see Making a boring bar holder on page 48) can be used as the basis of a simple tailstock; all that is needed is to cut a length of steel rod of a diameter that matches one of the collets, and machine a 60-degree taper on the end of the rod.

Putting it all together

This completes the constructional details. The remaining paragraphs describe how the components can be combined to provide the different indexing functions. The photos used show the division components used with the dividing head spindle, but all setups illustrated can just as easily be achieved with the division components attached to the lathe headstock itself.

Direct dividing setup

Figure 110 - Direct division

A set-up for direct dividing can be seen in Figure 110. The support bracket is bolted in place on top of the spindle, and a length of 5/16" steel bar is used as a support for the detent. The complete indexing arm assembly, including the boss, is used as the detent; the adjustable clamp on the support bracket allows the detent pin to be aligned with the selected hole circle – in this case, one of the circles I drilled in the 30T worm wheel. Using this arrangement, it is highly advisable to make use of the spindle brake to lock the spindle between operations; otherwise, all cutting forces will bear directly on the detent arm and pin, which are not designed to take heavy loads. The situation is rather different when the worm drive is in use, as the worm and wheel will naturally resist any movement of the output shaft.

Simple and compound dividing setup

Figure 111 - Compound division setup

Figure 111 (and also Figure 106 earlier) show the dividing head configured for compound division, for use on the lathe cross slide with the spindle pointing to the left. Reversing the "handedness" is achieved as follows. Remove the dividing assembly from the support bracket and reverse the fitting of the brake pinch bolt assembly. Remove the secondary worm assembly by loosening its pinch bolt, flip it over and re-fit it from the other side; unscrew the thumbscrew and re-fit it in the other end of the worm lock. Loosen the pinch bolt holding the 5/16" support bar into the primary worm carrier and slide it through so that it now protrudes from the other side; re-tighten the pinch bolt. The dividing assembly can now be re-fitted to the support bracket, using the other of the two split clamps.

Simple division

For simple division, ignore the secondary worm; lock it in position using the thumb screw, or remove it and fit the pins to lock the indexing plate boss in position. Select the desired hole circle, adjusting the indexing arm clamp to line up the indexing pin correctly. Calculate how many complete revolutions, plus how many holes in the circle, are needed between each division, and adjust the sector arms to expose the requisite number of holes in the circle. This calculation goes as follows:

— Multiply the number of holes in the circle by the ratio of the worm drive. For example, with a 20-hole circle, this gives 600.

- Divide the result by the required number of divisions. For example, to achieve 12 divisions with this combination, 600/12 = 50.
- Divide the resultant number by the drive ratio to give an integer result (which may be zero) and a remainder. The integer result (1 in this example) tells you how many full turns of the indexing arm; the remainder (20 in this example) tells you how many holes in the hole circle are needed in addition to the number of full turns. Set the sector arms to expose one more than the number of holes given by the remainder.

Compound division

The dividing attachment is assembled with the secondary worm drive in place, the locking pins removed, and a plate with a 30-hole circle fitted to the indexing plate boss. With this arrangement, the dividing attachment becomes a means of positioning the output spindle in increments of 1/100th of a degree, corresponding to a single division on the graduated thimble. Hence, a full rotation of the graduated thimble (40 divisions) gives 0.4 degrees of spindle rotation; this is exactly the same as moving the indexing arm by one hole in the 30-hole circle. A full rotation of the indexing arm gives 12 degrees on the spindle; 30 rotations of the indexing arm gives a full 360 degrees. In order to index between divisions, it is likely to be necessary to move both the indexing arm and the thimble.

To calculate the necessary movements, proceed as follows:
- Divide 360 by the number of divisions required giving a (probably fractional) number of degrees per division.
- Divide the number of degrees per division by 12 to determine the number of full turns of the indexing arm per division (which will be 0 if the number of degrees per division is less than 12), and a remainder.
- Divide the remainder by 0.4 to determine the number of holes in the 30-hole circle per division, and a remainder.
- The remainder tells you how many hundredths of a degree to move the thimble per division.

For example, creating 16 equal divisions requires 22.5 degrees per division. This is equivalent to one full rotation of the indexing arm (12 degrees), plus 26 holes in the 30-hole circle (10.4 degrees) plus 10 divisions on the thimble (0.1 degrees), giving a total of 22.5.

Where the calculation does not result in an integral number of 1/100ths of a degree per division, an allowance needs to be made for the error in calculating the movements per division by adding in an extra 1/100th of a degree every so often, so that the error is evenly spread out among the divisions rather than being "concentrated" in the last division. To give a very simple example, if you wanted to produce 320 divisions, then it would be necessary to rotate the spindle by 1.125 degrees per division. This is equivalent to 2 holes in the 30-hole circle, plus 32 divisions on the thimble, giving 1.12 degrees. However, this is 0.005 of a degree too little; on all but the last division, this would not be noticeable, but the error in the size of the last division would be 320 times the individual error, or 1.6 degrees. The last division would end up more than twice the desired width. The solution is to use 32/100ths on the thimble for the first, third, fifth,...and 319th division, and use 33/100ths for the even numbered divisions.

Keeping track of the next setting of the thimble can be tricky, especially if the number is changing from division to division. The safest approach is to calculate it all out beforehand, and to make up a table detailing where you expect to move the indexing arm and thimble to create each division. The alternative is to work it out in your head...and run the risk of messing up your evening's work.

Using a protractor

For many purposes, a protractor can be used as a simpler alternative to compound division. The protractor is in effect a 360-hole circle, giving a means of positioning the output spindle to a resolution of 1/30ths of a degree. Either way, the calculations involved are similar to those described above; again, it is well worth writing out a table of protractor positions for each division in order to avoid turning your hard work into scrap by careless application of mental arithmetic!

Chapter Ten
Adding a leadscrew

One of the limitations of the Taig lathe, as the uses to which it is put become more sophisticated, is that it has no leadscrew; hence, it cannot be configured to allow it to be used for cutting threads. There is an optional power feed that is available for the Taig lathe (see *Power feed* on page 76), but this is not able to be used for screw cutting as there is no facility for varying the drive ratio between the spindle and the leadscrew. The power feed is of use only as a powered "fine feed" for use when turning.

Other lathes of a similar size have a leadscrew with screw cutting ability; notable among these is the Sherline lathe (see *Other suppliers* on page 186), that also allows fine feed to be obtained via a supplementary motor attached to its leadscrew. An accessory kit is also available for the Sherline that adds the screw cutting capability; this consists of a selection of change wheels, sector arms, and a handwheel that allows the lathe spindle to be driven by hand during the screw cutting process, rather than by the lathe motor.

The thread cutting capability of the Sherline kit is fairly comprehensive. The Imperial version of the Sherline lathe has a 20 TPI leadscrew, and the kit allows most of the useful Imperial threads to be cut. The kit includes a 127-tooth wheel to allow Metric threads to be cut as well as Imperial threads. This introduces one of the curiosities of the Sherline change wheel set. The majority of the wheels are 24DP, however, there are a small number of 60DP wheels; two 100t wheels, one 127T and one 50T. These are used exclusively as the first driver and driven wheels in the Sherline set-up, using the smaller tooth pitch in order to include the magic 127T wheel for metric conversion without the wheel size being ridiculously large. The 24DP wheels are bored 3/8"; the 60DP wheels are bored 9/16".

At this point, the reader might reasonably be asking why the author is going on about a Sherline accessory in a book about the Taig lathe. The point is that the Sherline threading kit provides a very useful starting point for building threading capability into a Taig lathe; in particular, it is a cheap and convenient way of obtaining a set of change wheels of a suitable size, as well as other components that will prove useful.

It was a design goal to improve on the "hand cranked" approach of the Sherline threading kit, and have the drive derived from the spindle motor, as this would allow provision of a fine feed capability. Including a tumbler reverse was also considered desirable, to allow forward/reverse/neutral drive to the leadscrew. Similarly, incorporating a "dog clutch" to allow the leadscrew to be disengaged from the gear train; otherwise, fine feeding the carriage by hand becomes a little strenuous[15]. This will be apparent to anyone who has tried hand-feeding the carriage on a lathe with no clutch, with a fine feed gear train engaged; the friction generated by the fine feed gear train being back-driven is considerable. Also, the ability to disengage the saddle from the leadscrew (not possible on the Sherline) is highly desirable, thus allowing the Taig's existing rack & pinion to be used for fast saddle traverse.

Attaching the drive to the spindle

The Taig lathe uses a stepped pulley as its primary drive from a similar stepped pulley on the motor shaft. There is very little space between the pulley and the headstock, and the pulley overhangs beyond the end of the lathe spindle, so direct attachment of a gear wheel to the spindle itself is not a good option. A close examination of the pulley itself revealed that there is just about enough metal at the tail end of the pulley (the smallest diameter end) to make it possible to attach a small gear wheel directly to the body of the pulley using longitudinal screws through the gear wheel, threaded into the body of the pulley itself.

The Sherline gears are aluminium, and around 3/16" thick; while these might be OK for light, occasional use, something more robust would be needed as the basis for the tumbler reverse arrangement and for the gears used in the fine feed configuration. All of the additional gears used in this design are chosen from the 24DP range manufactured by HPC Gears (see Other suppliers on page 186). These gears are 5/16" thick; bore diameters vary according to the wheel size. The ones used in this design are steel; however, HPC offer the same range in Delrin, which, while I

---

[15] I have a Myford ML7 lathe; in standard form, it has a tumbler reverse, but no dog clutch, so in some respects, this design is superior to that of its big brother.

have not attempted to try them, may well prove to be a viable and cheaper alternative.

The primary drive wheel attached to the spindle is a 20T 24DP wheel, HPC part number G-24-20-PG. The 20T wheel has a 5/16" bore; this needs to be opened out to the same diameter as the pulley bore, nominally 3/8", in order for the lathe spindle bore not to be obstructed. Opening out the bore of this wheel retains the ability to fit the standard drill chuck spindle to the headstock, and also to use the full diameter of the spindle bore when turning small diameter stock.

It is straightforward to hold these gears in the 4-jaw chuck. By locating the tapered ends of the four jaws in the gaps between opposing pairs of teeth, the wheel can be "automatically" centred to a workable degree of accuracy, assuming that the 4-jaw is itself accurate. Obviously, if inspection with a dial indicator shows that the wheel is not properly centred, this can be remedied using packing shims between the jaws and the wheel as necessary. The bore can then be opened out using a suitable boring bar; in my case, I used a 1/4" boring bar fitted with a Sumitomo titanium carbide insert, which cuts very nicely indeed at these diameters. Figure 112 shows one of the tumbler gears having its bore modified by this means.

Figure 112 - Modifying a wheel bore

At first sight, this technique for opening the bore of a gear wheel sounds as if it will damage the gear teeth. However, the wheel is very well located by the tips of the four jaws and therefore only needs to be held with light pressure. A boring bar of this size cannot take heavy cuts; so not very much force is applied to the teeth in this machining operation, and considerably less than they are designed to handle. My own gears showed no evidence of tooth damage after being modified in this way.

The groove of the smallest drive pulley is approximately 0.66" diameter at the base of the V; this means that there is a wall thickness of approximately 0.14" to play with, which is enough thickness to take an 8BA screw. While the wheel is still in the 4-jaw chuck, it is a simple matter to scribe a circle at a diameter of 0.51" on the face of the wheel. Then mark the four mounting hole positions on this circle, 90 degrees apart (use the "prop" technique described in *Simple dividing operations* on page 108). The pulley is then removed from the lathe spindle so that the wheel and pulley can be drilled in a single operation. I found that the wheel and pulley could be held in position during drilling with the aid of a 3/8" steel dowel and the use of Superglue as a temporary adhesive. Drill through the wheel into the end of the pulley using an 8BA tapping drill, then open out the holes in the gear wheel to 2.2mm to clear the screws. Disassembly of the wheel from the pulley can be achieved by drifting the dowel out and splitting the components apart after drilling; if you avoid cleaning the parts too thoroughly before Superglue application this helps the disassembly process! The application of a little heat can also help by softening the Superglue. Cleaning the glue residue off after disassembly completes the drilling operation.

Figure 113 - Primary drive gear wheel

The four holes in the pulley are then tapped 8BA, allowing the gear wheel to be fixed in position using 8BA screws. I used cheese-headed screws and found that the heads were just slightly too large to clear the 3/8" bore, so turned them down to a more suitable diameter. Accuracy of assembly of the wheel to the pulley can be ensured by re-insertion of the 3/8" dowel prior to tightening the screws, ensuring that the bores of the pulley and wheel are concentric. The end result can be seen in Figure 113, which also shows some of the partly machined tumbler support bracket components in position.

At this point, it is possible to re-fit the pulley to the lathe, which is just as well, as many of the subsequent operations require the lathe to be operational.

The tumbler support bracket

Figure 114 - Tumbler reverse

Figure 114 shows how the tumbler reverse is assembled onto the lathe. Construction makes use of the fact that the lathe headstock has longitudinal T-slots running along its base at each side; these provide a temporary fixing method that allows the position of the mounting bracket to be adjusted before final fixing. The mounting bracket is a simple U-shaped frame attached to the headstock; this will hold the tumbler gears in position below the drive gear that is now firmly attached to the lathe's drive pulley.

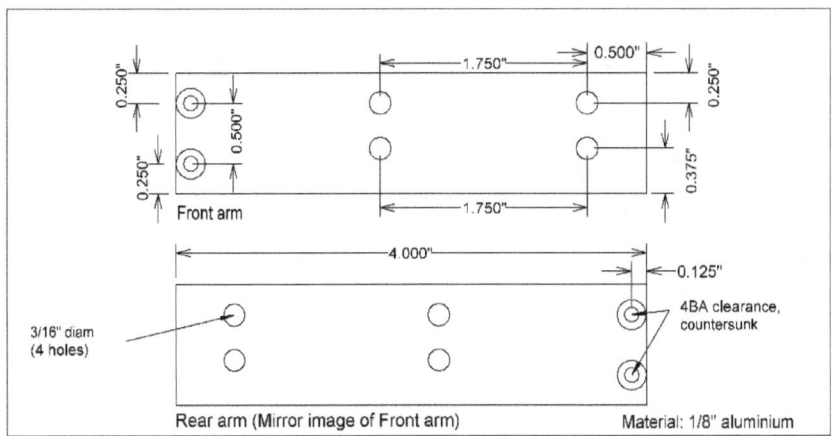

Figure 115 - Longitudinal arms

Construction of the tumbler support bracket is straightforward. The two longitudinal arms and the rear plate components are shown in Figure 115-Figure 117. The front and rear arms are mirror images of each other (Figure 115); both are made from 4" X 1" X 1/8" aluminium sheet, although the length may be adjusted during final fitting. Drill and countersink the 4BA clearance holes, and the two lower 3/16" holes in each plate; at present, do not drill the upper 3/16" holes. The lower holes are used to take T-bolts and nuts that will slot into the lower T slots on the headstock; 3/16" Whitworth, or 2BA, will work quite well for this as an alternative to the Taig "standard" of UNF 10-32.

The back plate of the support bracket (Figure 116) is made from a 3¼" X 2" rectangle of ¼" aluminium plate, with one corner and some of the top edge sawn off as shown. This material is removed to ensure that there is no chance of the bracket fouling the drive belt or pulley when the unit is finally assembled.

The 3¼" dimension of the back plate should be adjusted to the exact width of the headstock base. Each end of the plate is drilled and tapped 4BA to take the screws to attach the front and rear arms of the bracket. An additional tumbler stop plate (Figure 117), formed from a rectangle 2" X ¾" of ¼" thick aluminium is attached at the right-hand side of the back plate (as viewed from the rear of the lathe) by means of two 2BA countersunk screws. This stop plate will eventually be turned to an arc on its outer edge, centred on the tumbler pivot pin, as shown by the dashed arcs in Figure 116 and Figure 117, but not before the remainder of the tumbler has been constructed and test fitted in its final position. Similarly, the ¼" mounting hole for the tumbler pivot pin is not drilled until later.

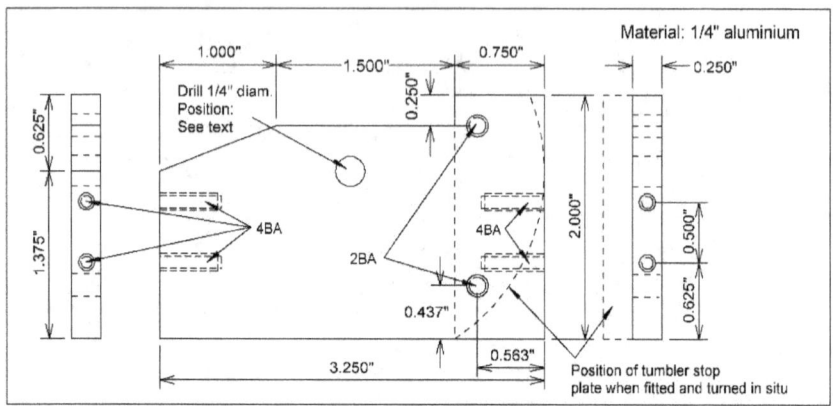

Figure 116 – Back plate

The two arms can now be assembled with the back plate, the additional rectangle for the stop plate screwed in place, and the whole bracket temporarily fitted to the headstock by means of T-bolts through the lower pairs of 3/16" holes.

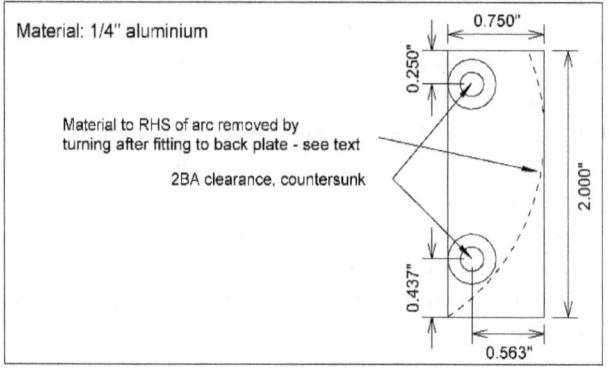

Figure 117 – Tumbler stop plate

The tumbler assembly

The tumbler assembly is carried on a ¼" thick brass Tumbler Plate, as seen in Figure 118. As the drive wheel attached to the spindle is 20T, the final drive gear from the tumbler reverse is also 20T; the two intermediate wheels are 18T (HPC part number G-24-18-PG) and 12T (HPC part number G-24-12-PG). The difference in diameters between the intermediate wheels is rather greater than is strictly necessary for operation of the tumbler, so a larger wheel might be substituted for the 12T wheel if desired, with suitable dimension changes elsewhere.

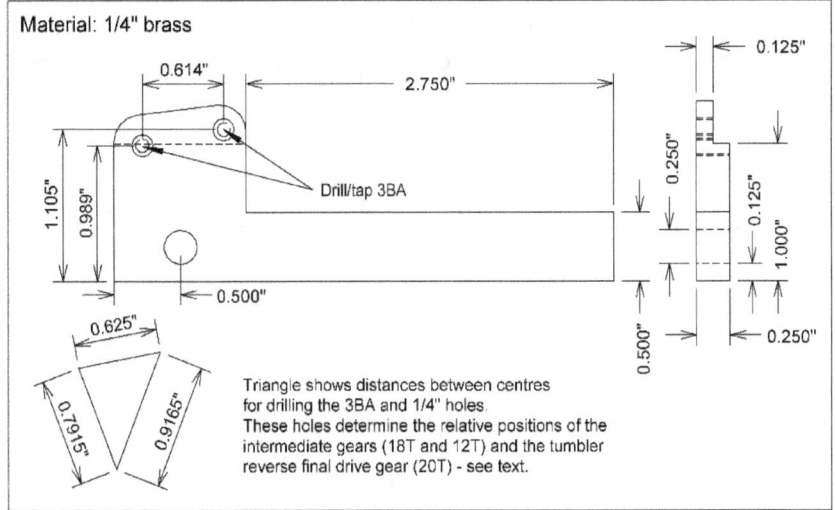

Figure 118 - Tumbler plate

The dimensions shown for the relative positions of the gears are based on the Pitch Circle Diameters (PCDs) quoted by HPC for the gears used and should result in free running of the gears. Clearly, for safety, it is advisable to check that these dimensions will be appropriate for the particular gears used and adjust if necessary. The calculation is simple; the 0.7915" dimension is half the sum of the PCDs of the 20T and 18T wheels. The 0.625" dimension is half the sum of the PCDs of the 18T and 12T wheels. Finally, the 0.9165" dimension is half the PCD of the 20T wheel, plus the PCD of the 18T wheel, less half the PCD of the 12T wheel.

The two 3BA holes will hold the intermediate wheel studs; the ¼" hole will fit over the pivot stud for the tumbler, which also carries the tumbler output gear and the first driver wheel of the change wheel train.

The plate is cut to size and the three holes drilled and tapped as indicated; then the recess on the rear face is milled or filed out. The recess at the back is necessary in order to avoid the plate fouling the lathe drive pulley. An alternative to using ¼" sheet is to fabricate the plate from two pieces of 1/8" thick sheet, soft soldered together. An advantage of the latter approach is that the hole positions for either of the two 3BA holes can be adjusted more easily if they turn out to be incorrect for the gears to mesh nicely. Remove part of the front face by cutting to 1/8" depth with a hacksaw and unsoldering it from the rest of the plate. It is then a simple matter to solder in a fresh piece of 1/8" thick brass in its place and re-drill/tap the hole in the correct position. It is advisable to make the arm of the tumbler plate slightly over length to start with, and to adjust it later as described below.

Two steel studs are turned to carry the 18T and 12T wheels. The 18T stud (Figure 119) is turned from ½" diameter steel rod; the 18T wheel has a 5/16" bore, so the stud is turned down to this diameter, leaving sufficient length to be further turned down and threaded 3BA for the last 3/16" of its length.

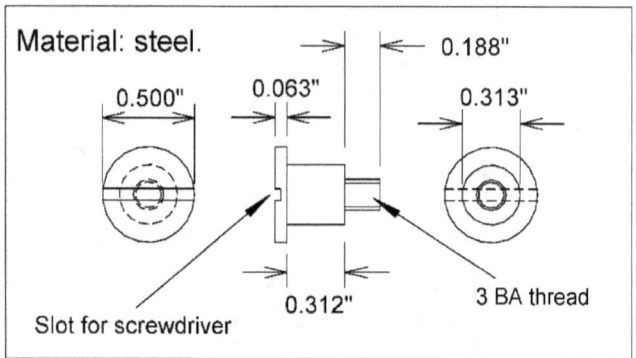

Figure 119 - 18T stud

The 5/16" diameter portion is nominally 5/16" long to match the thickness of the 18T wheel; however, this should be left a few thou over length to ensure that the wheel will run freely when the stud is screwed onto the tumbler plate. The stud is parted off to leave a screw head about 1/16" thick, and a screwdriver slot is cut across the head with a hacksaw. The 12T stud (Figure 120) follows the same procedure, this time starting with 3/8" stock and reducing to ¼" for the bore size of the 12T wheel.

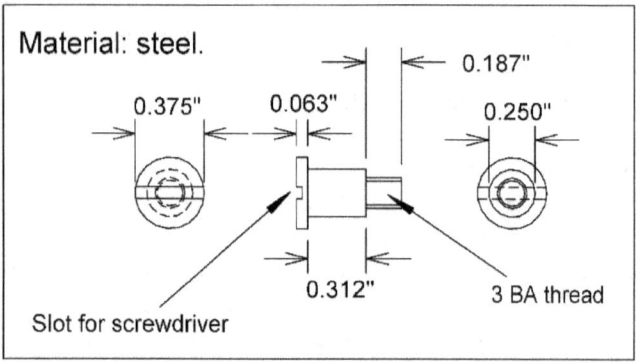

Figure 120 - 12T stud

The two wheels can now be attached to the tumbler plate, with a suitable diameter, thin steel washer under each wheel stud to prevent direct contact between the wheels and the tumbler plate. Note that the 12T stud is fitted in the hole furthest from the ¼" diameter hole. If all is well, the two wheels should mesh nicely and rotate freely.

The arm of the tumbler plate carries a locking plunger attached at its far end. This plunger will eventually locate in three holes in the curved edge of the tumbler stop plate, giving locking positions for Forward, Neutral and Reverse for the tumbler. The plunger runs in the Tumbler Lock Body, a block of ¼" x 1/2" brass (Figure 121).

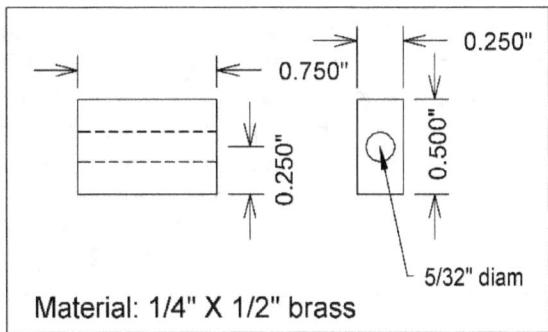

Figure 121 - Tumbler lock body

The Tumbler Lock Body is soft soldered onto the end of the operating arm, on the side that has the recess at the top edge. It is aligned so that the axial hole points along the centre line of the arm; i.e., on a line to the centre of the ¼" diameter hole in the tumbler plate. However, this is not soldered in position until the radius is cut on the stop plate as described below. Once soldered in place, any excess length of the operating arm is trimmed off flush with the end of the tumbler lock body, and the locking plunger fitted. The locking plunger is a length of 5/32" diameter silver steel rod (Figure 122), turned to a radius at one end. Silver steel (as it is known in the UK) is what is called "Drill rod" in the USA, a carbon steel rod, usually centerless ground, that can be oil or water hardened.

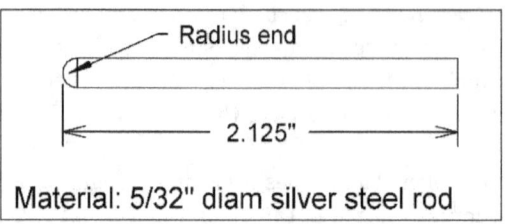

Figure 122 - Tumbler lock plunger

A brass knob cut from 3/8" diameter rod (Figure 123) is attached at the square end using Superglue, and a thin brass washer cut from 1/4" brass rod (Figure 123) acts as the retainer for the return spring at the radiused end. The knob is glued in place, the radiused end is passed through the axial hole in the lock body, a length of spring from a cheap ballpoint pen serves as a return spring, and the retaining washer is Superglued in place about ½ way along the exposed length of the plunger. Again, the final gluing of the retaining washer is left until the lock body is soldered in place and the arm has been trimmed to length.

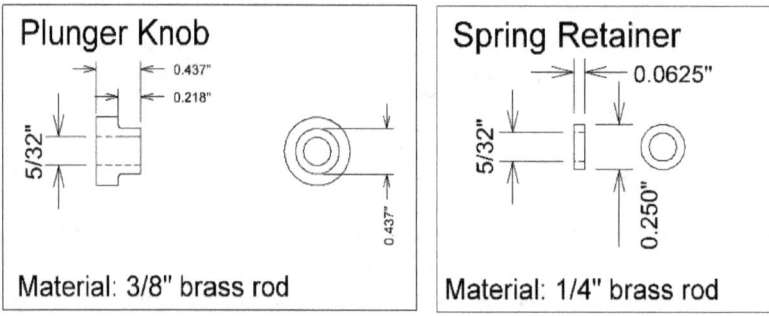

Figure 123 - Plunger knob and spring retainer

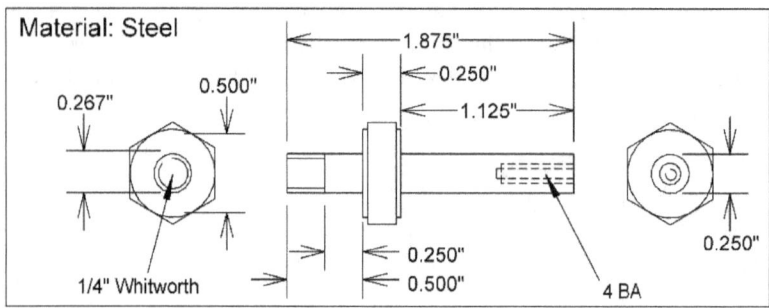

Figure 124 - Tumbler pivot pin

The Tumbler Pivot Pin (Figure 124) is cut from a length of hexagonal steel stock, nominally 0.56" across flats in the sample I used, but this is not critical. Face one end and reduce its diameter to ¼" over 1.25" of its length. This length will need to be adjusted down later, once the rest of the assembly is complete; for the moment leave it over length. Reduce the diameter to 0.56" for about 1/32" in order to present a circular shoulder. Drill and tap axially 4BA for the screw that will act as the retainer for the final drive boss and the tumbler plate. Reverse the component, face the end, and reduce the diameter to ¼", leaving a shoulder ¼" wide, again turning off the hexagon for about 1/32". Thread this end ¼ Whitworth and reduce the threaded portion to about ½" long. This acts as the mounting stud for the pivot pin and will be fitted through the ¼" hole in the support bracket back plate.

The tumbler final drive gear (a 20T gear, see above) is carried on a bobbin that also carries the first driver gear wheel of the change wheel gear train. The Sherline 24DP gears have a 1/8" wide keyway cut into the bore; the bobbin has an integral key to match, hence allowing any of the Sherline 24DP gears to be used, unmodified, as the first gear in the change wheel gear train. The 60DP gears can also be used once modified to reduce the bore to 3/8", as described later in this article. Construction of the bobbin starts by increasing the bore size of a second 20T gear (HPC part number G-24-20-PG) to ½" diameter, using the technique already described, illustrated in Figure 112. The bobbin is shown in Figure 125.

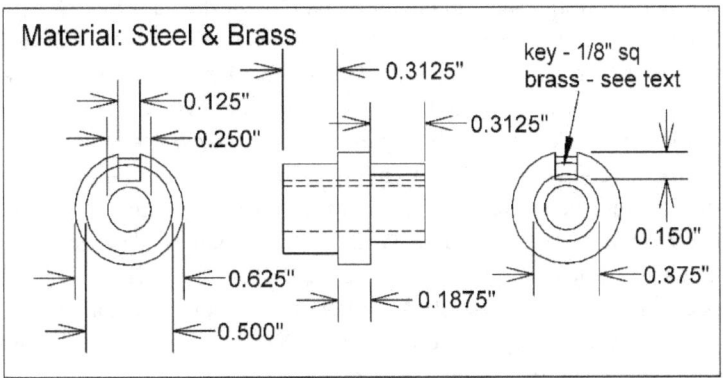

Figure 125 - Tumbler final drive bobbin

Chuck a suitable length of 5/8" round steel bar; turn down the last 5/16" to be a press fit into the bore of the 20T wheel. Leave a 3/16" shoulder at full diameter, then turn down a further 5/16" length to 3/8" diameter, bore the piece ¼" diameter and part off to length. The channel for the keyway is milled (or very carefully filed) to 1/8" wide along the entire length of the part, taking care not to cut too deeply into the 3/8" diameter portion. Milling is to be preferred; the Taig milling slide is invaluable for this kind of operation. A length of 1/8" square brass bar is then cut and soft soldered into the milled channel. Return the part to the lathe to clean up the ends of the key, and also to trim the excess that will protrude from the ½" diameter end. Press the 20T gear into place, securing with a suitable adhesive if necessary. The remainder of the brass key may need some final adjustment to fit the keyways in the Sherline gears. Leaving the keyed portion of the bobbin at 5/16" length will allow either Sherline gears or further HPC 24DP gears to be used in the change wheel train.

Now that all the components for the tumbler assembly have been made, the final machining and setting up of the support bracket and the tumbler can be completed. First, machine the pivot pin to its final length. Place the tumbler plate over the pivot, followed by a suitable washer (of the same thickness as those used under the idler wheels), and the completed final drive bobbin. The pivot pin should just slightly protrude from the bobbin; i.e., when a 4BA screw and washer is screwed into the end, it will retain the components in place while allowing the bobbin to run freely.

With the bobbin and plate assembled on the pin, check that the lower three gears of the tumbler now mesh correctly. Assuming all is well, mark the position of the hole in the back plate of the support bracket that will take the pivot pin. This should be vertically below the centre of the lathe spindle and should be positioned to allow the tumbler's idler wheels to mesh correctly with the pulley's drive gear. The position of the support bracket should be adjusted on its T bolts to ensure that all the gear wheels end up in the same plane. Drill the hole in the back plate ¼" diameter, fit the pivot pin in the hole with a suitable nut and washer, and check that the gears still mesh correctly in both forward and reverse positions. If necessary, the hole in the back plate can be enlarged slightly to get proper alignment.

Remove the back plate from the support bracket, screw the stop plate in place, and mount the assembly in the 4-jaw chuck to turn the arc on the stop plate. It will be necessary to reverse two of the jaws to achieve this feat; the plate is centred so that the pivot pin is on the centre line. While the back plate is removed, spot through the upper pairs of holes in the side arms of the bracket, drill and tap into the headstock extrusions for suitable 3/16" (or UNF 10-32) bolts to act as the permanent fixing method for the bracket. The temporary T bolts can then be removed, thus allowing the saddle stop to be used once more.

Re-assemble the back plate and pivot pin in position, checking that alignment is still good. Re-fit the operating arm to the tumbler plate; assemble it onto the pivot, followed by the washer and bobbin. Retain the components in place with a 4BA screw and washer (if you are feeling very energetic, make up some 4BA screws with suitably large heads as an alternative that avoids the perennial vanishing washer problem!). Temporarily clamp the locking assembly in position on the tumbler plate, adjusting the position of the lock body so that, when the locking pin is fully retracted, it just clears the edge of the stop plate. Remove the tumbler plate from the assembly, soft solder the lock body in place, finish the arm to length, re-fit the locking pin, spring and retainer, and fit the complete tumbler plate and locking assembly back onto the pivot. Mark on the stop plate the position of the tumbler locking pin when the tumbler is in the forward, neutral (disengaged) and reverse positions, and drill a centre hole at each position with a centre bit in preparation for drilling the locating holes for the locking pin. Drill locating holes at each of the three positions. These are ideally 5/32" diameter, and deep enough to allow the locking pin to extend fully into the locating hole. If your drilling is accurate (i.e., exactly on a radius centred on the tumbler pivot pin), then 5/32 may work fine; for lesser mortals, enlarging these holes to 3/16" or 7/32" may be necessary for easy operation of the tumbler lock.

At this point, you should have an operational tumbler reverse drive in position, capable of taking gearwheels of up to 5/16" in thickness, without disturbing the original functions of the lathe. One minor detail here is that the steel rod used as the saddle stop will now be too long for operations close to the chuck, as it fouls the support bracket's back plate. This can be remedied by cutting a second, shorter, saddle stop bar to be used under these circumstances, retaining the original one for working further from the chuck.

The dog clutch

Figure 126 shows the general positioning of the dog clutch body (Figure 127); the clutch is supported on a simple aluminium angle bracket that is bolted into the aluminium foot of the lathe. The clutch mounting bracket also provides a convenient retaining cover for the operating components of the clutch. The brass operating lever can be seen protruding from the clutch body, in the disengaged position.

Figure 126 - The dog clutch in position

The clutch operation is very simple indeed. The shaft into the clutch (Figure 128) has a groove cut in it into which a 2BA bolt (seen on top of the clutch body) locates, thus ensuring that the shaft stays in its correct lateral position but can rotate freely. The visible end of the shaft is a bobbin similar to the one used for the output gear of the tumbler assembly (Figure 129); the main difference is its length and the fact that in this case it is an integral part of the shaft. A gear wheel can be placed on the bobbin, along with spacers to line it up with its driver wheel; another 4BA screw and washer retain these in position.

The opposite end of the input shaft (the end within the clutch body) is filed to a D section over the last ½" of its length. This D section constantly engages a short sliding shaft (Figure 130), which has a similar ½" D section at its near end, and a shorter, 3/8" long D section at the other. A groove in the centre of this shaft carries the clutch operating shoe (Figure 133), which in turn is pivoted in a hole in the clutch operating lever (Figure 132). Hence, as the operating lever moves to the left or right, so does the sliding shaft. The output shaft (Figure 131) from the clutch has a 3/8" long D section at its near end. The lengths of the components have been chosen such that, regardless of the position of the operating lever, the sliding shaft's D section always engages the D section of the input shaft. However, the D section of the output shaft is engaged only when the lever (and the sliding shaft) is moved to the right. This is crude but effective, and above all, fairly simple to make.

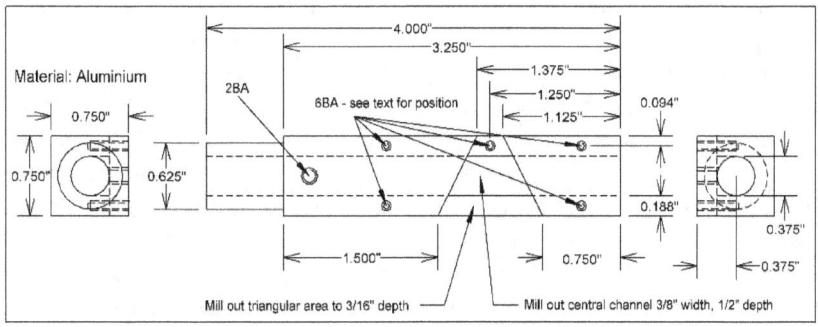

Figure 127 - Dog clutch body

The output shaft from the clutch is threaded 3/8" BSF (20 TPI) to take a suitable length of BSF studding for the leadscrew[16]. Clearly, this means that the "sense" of operation is the reverse of most leadscrews (conventionally, leadscrews use a left-hand thread); if this is a problem, it may be possible to obtain left-handed BSF studding (or have it cut) and left-handed taps. You will have detected by now that this did not worry me sufficiently to do anything about it!

Construction starts with the clutch body (Figure 127); this is made from a 4" length of 3/4" square aluminium.

Centre it accurately in the 4-jaw chuck, face off the ends to length, and turn down 3/4" of its length to a diameter of 5/8". This cylindrical section will carry the Sherline "banjo" parts, after modification as described below. It is then necessary to bore the entire length to 3/8" diameter. This is a non-trivial exercise in the Taig lathe, but it can be done, very carefully and with lots of lubrication, as the prototype proves. Alternatively, careful set-up on a stand drill might prove more satisfactory; it is not essential that the bore is accurately concentric with the 5/8" diameter portion. Mark out, drill and tap the 2BA hole that will carry the input shaft retaining screw. On the same face, mark out and mill the triangular shaped recess for the operating lever to slightly deeper than the thickness of the lever (nominally 3/16") to ensure that it will operate easily when the clutch body is screwed onto its bracket. You know when the right depth has been reached, as the cutter will just start to

---

[16] The design is based around a 20 TPI leadscrew in order to keep the thread cutting capability essentially the same as the Sherline lathe. As 3/8" BSF studding, taps etc. may not be readily available outside the UK, it may be appropriate for some readers to choose a different thread for the leadscrew. A suitable alternative in the USA would be 7/16"-20 UNF. There should be sufficient room in the design to accommodate the extra diameter without too much difficulty, but the details are left as an exercise for the reader!

break into the top of the bore (assuming that the bore is concentric!). The final milling operation is to open out the bore immediately below this recess to form a slot 3/8" wide to halfway across the bore (1/2" below the surface of the bar). This slot carries the clutch operating shoe, located in the groove in the sliding shaft. Drill and tap the 6 BA hole as marked; screw a short length of 6BA studding into the hole for the slot in the rear end of the operating lever to locate onto. Leave the remaining holes for the present; these are marked and drilled later on in conjunction with the mounting bracket.

Next, the input shaft. This is fabricated in two parts. The main body of the shaft (Figure 128) is made from 3/8" Silver Steel rod (chosen because of the surface finish, not for any other reason; BMS would do fine).

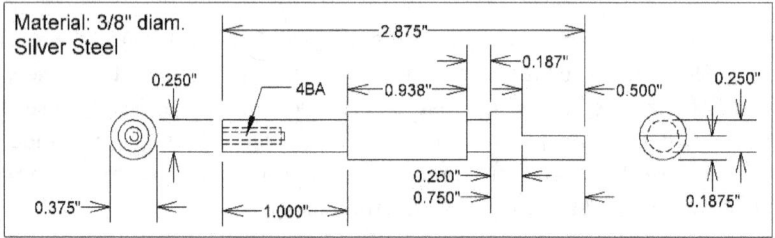

Figure 128 - Dog clutch input shaft

Face off a length to 2 7/8" long. Reduce the diameter over a 1" length to ¼", drill and tap 4BA for the gear retaining screw & washer. Cut the groove for the 2BA retaining screw; this groove starts at 15/16" in from the shoulder, 3/16" wide and reduces the diameter to ¼". Remove from the lathe, mark out and mill or file the ½" long flat at the thick end of the shaft. The second part is a bobbin to carry the gearwheels (Figure 129). Three of these are needed in all; one is fabricated and then soft soldered in place on the ¼" diameter portion of this input shaft, the other two are free running gear carriers for use on the intermediate studs in the gear train. Fabrication is similar to the tumbler output bobbin; turn down a length of ½" diameter stock to 3/8" diameter over a length of 13/16", axially drill ¼" diameter, and part off to leave a ½" diameter shoulder 3/16" long. The groove for the 1/8" square brass key is milled as before, and a 1" length of key soft soldered in position. Clean up the ends and any solder excess, return to the chuck and turn down the key to the ½" diameter; i.e., there should be 1/16" of key protruding above the surface when completed. Solder in place on the input shaft, taking care not to disturb the key in the process, and clean up again. Make the other two bobbins while you are at it!

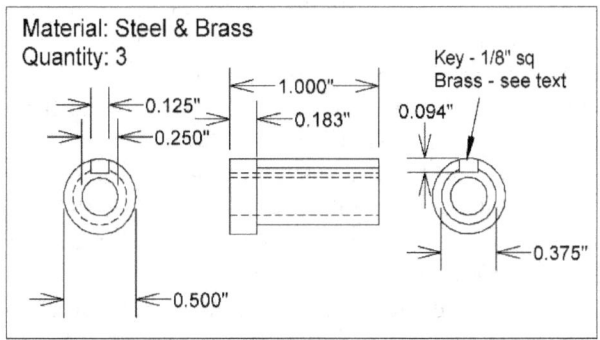

Figure 129 – Bobbins

The sliding shaft (Figure 130) is very straightforward; again, silver steel was used, but this time the groove is 3/8" wide, and there are two flats to mill/file.

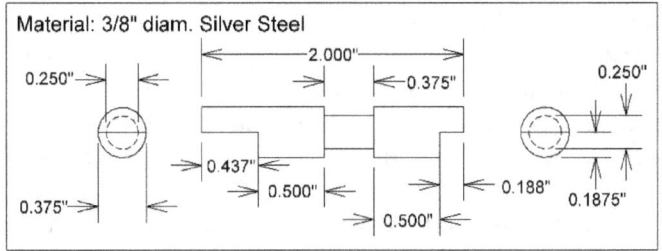

Figure 130 – Sliding shaft

The clutch output shaft (Figure 131) is slightly more involved. Chuck a suitable length of hex section steel bar; the one I used measured 0.56" across the flats, which is slightly smaller than a standard 3/8" BSF nut. Face off, axially drill and tap 3/8" BSF (20TPI) to a depth of 5/8".

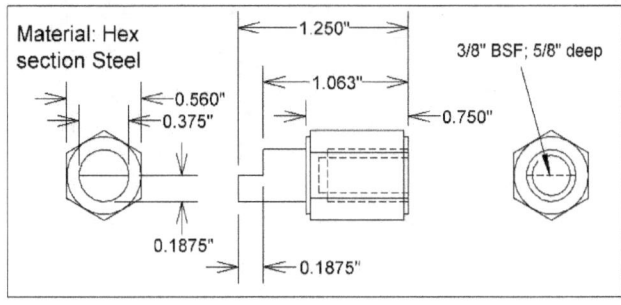

Figure 131 – Output shaft

Cut a small shoulder, about 1/32", to the inscribed circle diameter of the hexagon (0.56"). Remove from the chuck, screw the piece onto a short length of 3/8" BSF studding, and use this as a stub mandrel to hold the piece for the remaining machining operations. Mount the mandrel in the chuck so that the end of the hex rod buts against the chuck jaws; the turning action will tend to tighten it against the jaws. Turn down and face off to leave a ¾" long hex shoulder, and a ½" length of 3/8" diameter. File or mill the flat for the last 3/16" of its length. Using the studding as a mandrel in this way ensures that the 3/8" bearing diameter will be as near as possible to being concentric with the leadscrew.

The remaining components of the clutch are the shoe, the lever and the bracket. The lever (Figure 132) is straightforward; cut a 1 ¾" length of brass, 3/8" X 3/16". Radius the ends. Drill a ¼" hole through ½" from one end. Cut a 1/8" wide slot in the same end, to engage with the 6BA pivot pin in the clutch body.

The shoe (Figure 133) is cut from a length of 3/8" square section brass. Centre a 1" length in the 4-jaw chuck, turn down one end to ¼" diameter over a length of 3/16" (this fits in the hole in the lever).

Remove from the chuck and cross-drill the bar 3/8" from the end of the turned section. Carefully cut and file across this hole so as to leave a semi-circular bearing surface that will engage with the slot in the sliding shaft. File the two bevels; these are necessary to allow full left and right travel of the shoe in the slot in the clutch body.

Finally, cut a 2 ¾" length of 1"X1"X1/8" Aluminium angle to form the mounting bracket (Figure 134). Do not drill the four 6BA clearance holes yet; the positions for these are described in the next section.

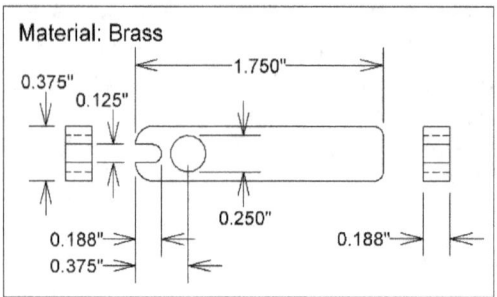

Figure 132 - Clutch operating lever

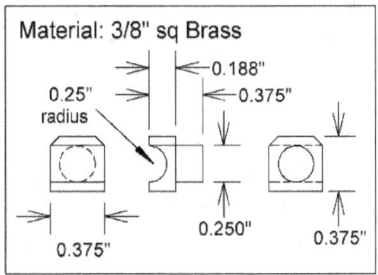

Figure 133 - Clutch operating shoe

The clutch can now be assembled and tested for correct operation. Insert the clutch components, fit the 2BA retaining screw, locking it with a nut so that the input shaft is captive but can rotate freely. Make sure that the clutch will engage and disengage correctly, adjusting if necessary.

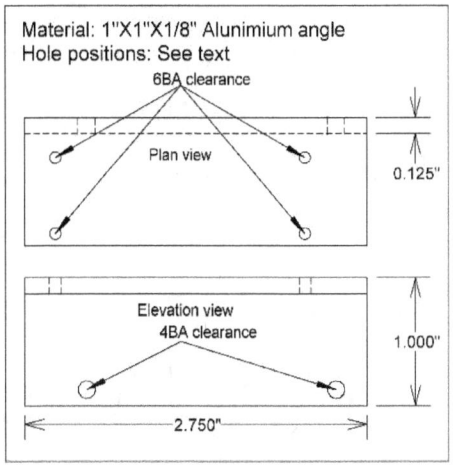

Figure 134 - Clutch mounting bracket

Fitting the clutch

The next stage is to fit the clutch to the lathe. The objective here is to end up with the axis parallel to the axis of the lathe, and the shoulder of the input shaft bobbin in the same plane as the corresponding shoulder of the tumbler output bobbin; i.e., if a gear were to be placed on each bobbin, butting against the shoulder, both wheels would be in the same plane. For reasons that will become apparent later, a secondary objective is to allow the input shaft bobbin to overhang any mounting board or bench, to make room for large gears. A word about lathe mounting is therefore appropriate at this point.

My lathe is mounted on the Peatol motor mount; this is a 10" X 13" steel base plate approximately 1/8" thick that also carries the motor hinge plate (see Figure 6). The foot of the lathe is placed in the near left-hand corner of this plate. In turn, the plate is screwed to a 14" X 13" piece of chipboard, nominally 18mm thick. The left-hand edge of the plate is about 1" in from the left-hand edge of the chipboard. In turn, the chipboard is screwed/glued to a 21" X 15" piece of 1" thick blockboard, leaving a 2" apron in front and about 7" to the right of the chipboard. The whole assembly is mounted on ¾" high rubber feet. See Figure 135 for a rough plan of the baseboard.

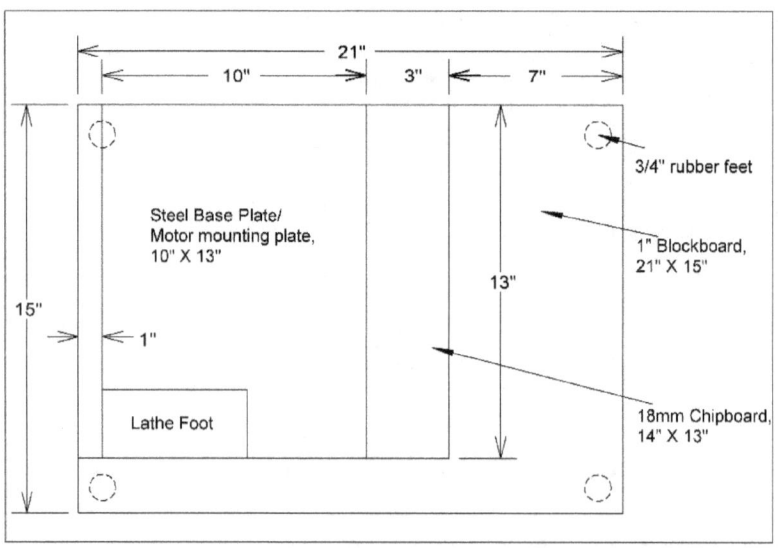

Figure 135 - Plan of baseboard

The net result is that the metal base plate for the lathe is raised by about 2.5" above the bench top on which the lathe sits. This gives ample overhang at the left hand side of the lathe to accommodate the change wheels, and provides a useful base for mounting the right hand end leadscrew bearing in a position that will allow the saddle to be removed without disassembly of the leadscrew components.

Mounting the clutch is a matter of careful measurement and marking out. Firstly, determine the lateral distance between the left-hand end of the lathe foot and the plane of the shoulder on the tumbler output bobbin. This gives the desired position of the input shaft bobbin's shoulder relative to the lathe foot. The support bracket should be placed so that its right-hand end will match the position of the right hand end of the clutch body. Place it in position, with its lower edge resting on the mounting plate under the foot of the lathe. Mark out and drill the 4BA clearance holes; spot through these onto the foot itself, drill and tap the foot 4BA. The bracket can now be screwed into position. The clutch body is then clamped temporarily to the bracket, leaving 1/8" clearance at the back of the clutch between it and the bracket, carefully checking its alignment relative to the tumbler output bobbin. Place register marks on the clutch and bracket, unscrew the bracket, clamp the clutch and bracket together so the register marks match, and drill clutch and bracket in one operation for the four 6BA mounting screws. Use the tapping drill to the full depth through both pieces; then at the same setting, open out to 6BA clearance through the bracket. This ensures that the components will re-assemble in the right place.

The bracket can then be replaced in its final position, and the clutch fitted to it with suitable screws. If desired, the free end of the bracket at the left-hand side can be supported further by means of a simple angle bracket screwed to the baseboard.

The leadscrew

Figure 136 gives an overall view of the leadscrew and its components. The leadscrew is a length of 3/8" BSF studding, giving 20TPI. One end screws into the threaded portion of the clutch output shaft; the other end has an end nut attached (shown in Figure 137), which is located in a simple aluminium bracket that carries the leadscrew thrust bearings. A handwheel is attached to the right-hand end of the leadscrew.

Figure 136 - Leadscrew layout

The studding chosen for the leadscrew should be as straight as possible. I managed to obtain two different samples, both of which proved to be slightly bent on close inspection. However, with careful straightening, I found that it was possible to produce a length of, say 18" long, that was straight enough so that any remaining inaccuracy was not visible to the eye on final assembly. The use of a flat surface, for example, a piece of float glass, helps a lot in determining how straight (or otherwise) the studding is, and in making any necessary corrections.

The actual length of studding required will depend upon the chosen layout; as indicated earlier, I chose to place the right-hand bracket sufficiently far beyond the end of the lathe bed so that it does not prevent removal of the saddle when the leadscrew is installed. An alternative would be to devise a way of attaching a bracket to the right-hand end of the lathe bed. If you attempt to do this, and the attachment method requires holes to be drilled in the lathe, it is worth remembering that the aluminium extrusion that supports the lathe bed is filled with a concrete-like material!

Construction of the end nut (Figure 137) is very similar to the technique described for the clutch output shaft. The same 0.56" AF Hex section steel was used, and the same procedure, using a short length of studding as a mandrel, employed to ensure that the plain-turned portion is near enough concentric with the threaded socket. The nut has an axial 2BA-threaded hole in it that will provide a means of adjusting end play in the thrust bearings.

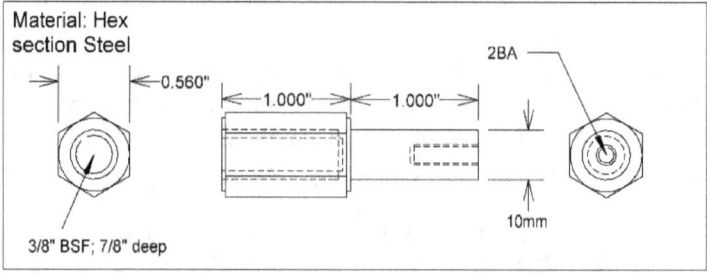

Figure 137 - Leadscrew end nut

The leadscrew bearing bracket is shown in Figure 138. It is fabricated from a 2" length of 2"X2"X1/4" aluminium angle, with two triangular supports (also shown in Figure 138) added to improve its rigidity. The supports are held in place using 4BA countersunk screws, and the base of the bracket is drilled for 3/8" bolts that will attach it to the baseboard of the lathe. The only difficulty in construction is deciding where to drill the hole for the bearing for the end nut. Careful measurement is required here, backed up with the use of washers under the bracket to adjust its height if necessary; the objective being to ensure that the leadscrew runs parallel to the lathe bed. The bearing itself consists of a ½" diameter plain brass or phosphor

THE TAIG/PEATOL LATHE

bronze bearing, bored 10mm diameter, inserted in the bracket and supported by a pair of needle roller thrust bearings, one placed either side of the bracket. The brass insert is a simple turning job from ½" diameter stock, and is a press fit in the bracket. The roller bearings are about 23mm in diameter, bored 10mm, and about 4mm thick; each bearing consists of three components, a roller cage and a pair of steel washers. These roller bearings can be obtained from RS Components, part number 198-8850.

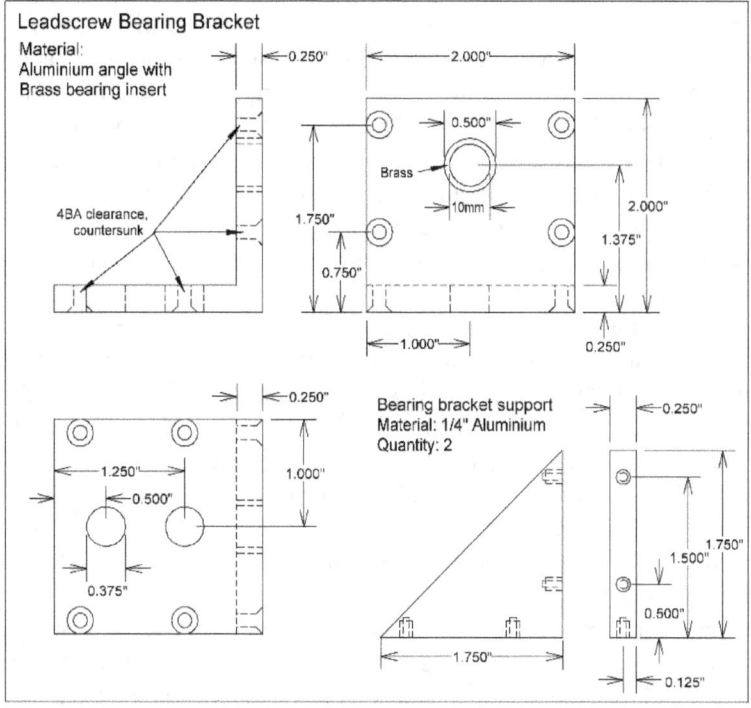

Figure 138 - Leadscrew bearing bracket and supports

The bearing arrangement described here may be a little excessive for most uses; a plain bearing will suffice for most applications, as used on many commercial leadscrews. However, the arrangement described works very well, and one of my guiding thoughts was that I might be tempted at some point to convert the lathe to CNC operation, where the ability to remove end play in the leadscrew would be useful.

When locating the bracket on the baseboard, care needs to be taken to ensure that the bearing is positioned such that the leadscrew will be parallel to the lathe bed when finally fitted. Drilling the mounting holes in the baseboard slightly oversize will help here, allowing some adjustment of the final position of the bracket; it may also be necessary to place packing shims under the bracket to adjust the height of the screw above the baseboard.

Once the bracket is mounted in place on the baseboard, the leadscrew can be cut to its final length. First, screw the clutch output shaft onto one end of the leadscrew, and soft solder it in place. Apply a little flux to the two parts and drop a couple of short lengths of solder wire into the threaded socket prior to screwing the two components together, heat the joint and apply more solder when the joint has heated up. An alternative would be to use Superglue as a retainer. Next, fit the end nut to the other end of the length of studding and measure the distance between the shoulders on the two end nuts. Now measure the distance between the thrust face of the leadscrew bracket and the end of the dog clutch body. The difference between these measurements, and a bit to allow for the thickness of one of the needle roller bearings, tells you how much to shorten the length of studding. Cut off the excess, screw a 3/8" BSF nut onto the end, followed by the leadscrew end nut. The BSF nut is used as a lock nut, allowing fine adjustments to the screw length to be made once all is properly aligned.

It is now possible to fit the leadscrew in its final position. Unscrew the clutch body from its bracket, fit one of the roller bearings over the end of the leadscrew, and then fit the end of the leadscrew into its bracket. Fit the dog clutch body over the other end of the leadscrew and screw it back in place on its bracket. The final position and height of the leadscrew bracket can now be adjusted, to ensure that the leadscrew will run parallel to the axis of the lathe in both the vertical and horizontal planes. Tighten the mounting bolts of the bracket, then adjust the end nut and its lock nut to remove all but a couple of thou of end play between the clutch output shaft and the clutch body.

The final leadscrew component is the handwheel. This serves two purposes; firstly, the obvious one of allowing the leadscrew to be turned by hand, and secondly, it provides the end play adjustment for the leadscrew thrust bearings.

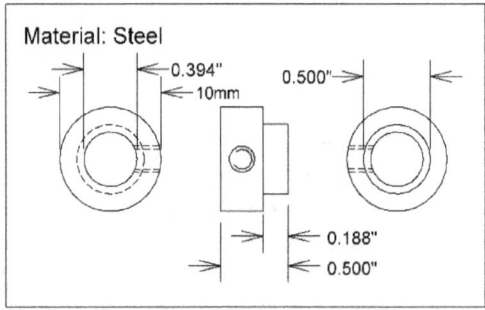

Figure 139 - Handwheel boss

The handwheel boss, Figure 139, is machined from ¾" diameter steel bar. A short length is faced both ends for a final length of ½", bored 10mm to fit the end of the leadscrew end nut, and 3/16" of its length turned down to ½" diameter. A radial hole is drilled and tapped 10-32 UNF to take a set screw.

The handwheel disc, Figure 140, is a 2.5" diameter brass disc, bored ½" at its centre to take the boss, and with a hole near the edge tapped 4BA to take the handle. The disc is soft soldered or Superglued onto the boss, the excess cleaned off and the handle fitted.

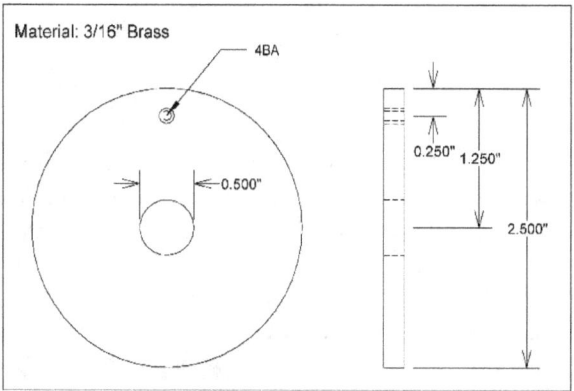

Figure 140 - Handwheel disc

The handle is simply a 5/8" length of brass rod, slightly waisted, and axially drilled 3.6mm to clear a 4BA domed head brass screw. The screw is passed through the handle, a 4BA nut Superglued in place so that the handle spins freely, and the completed handle is then screwed into place on the handwheel[17].

Any excess bolt length can then be cleaned off at the back of the handwheel. The final finishing touch to the handwheel is to mark 50 divisions on its circumference, corresponding to thousandths of an inch of travel of the saddle. This is relatively easy to do once the banjo components have been modified, allowing a train of change wheels to be attached to the spindle (alternatively, the techniques described in "Dividing and graduating" on page 93 could be used). Choosing suitable combinations of change wheels, it is possible to index the headstock, using a detent located between gear teeth to stop the rotation at each 1/50th of a revolution. A scriber mounted in the tool post is used to score the edge of the wheel; a short line for the intermediate marks and a full width line for every fifth mark. A set of small number punches can then be used to number every 5th mark. Clearly, you have a

[17] This handle is very similar to the modification to the existing Taig handles described in *Spinning handles* on page 79.

choice as to which way you number the marks, ascending numbers indicating left hand travel, or vice versa, remembering of course that if a conventional right-hand thread is used, the direction of saddle travel will be the reverse of the conventional "sense".

The second needle roller bearing and the completed handwheel are fitted to the end of the leadscrew, and any end float in the bearings is removed by means of a 2BA screw and washer screwed into the end of the end nut. The set screw fitted to the handwheel boss is then tightened to hold the wheel in place.

The split nut assembly

The arrangement described so far leaves 3/8" or so of clearance between the lower edge of the saddle casting and the top of the leadscrew. The split nut assembly is attached under the left-hand end of the saddle casting (see Figure 141). The attachment involves a certain amount of modification to the saddle, in order to provide a flat base for the split nut mounting plate, and also to allow an operating lever to protrude through the side of the casting.

Figure 141 - Split nut in situ

I must thank David Gingery for the idea behind the split nut design; the one described here operates in the same manner as the one he used in his lathe design, as described in his excellent series of books, "Building Your Own Metal Working Shop From Scrap"[18]. Instead of the more conventional clamping action found in most

[18] Gingery, David J, "Building Your Own Metal Working Shop From Scrap, Volume 2: The Metal Lathe", ISBN 0-9604330-1-5.

lathes, the split nuts in this design are offset from each other and are attached to a carrier that is rotated to disengage them from the leadscrew. The prototype literally used two halved 3/8" BSF nuts, offset by about ½" and soft soldered onto a piece of 1/16" steel plate to which was attached the operating shaft. This worked just fine but was a little crude and potentially not terribly robust, so the final "production" version was re-designed and involves machining the nut halves and carrier in one piece from ¾" square steel bar.

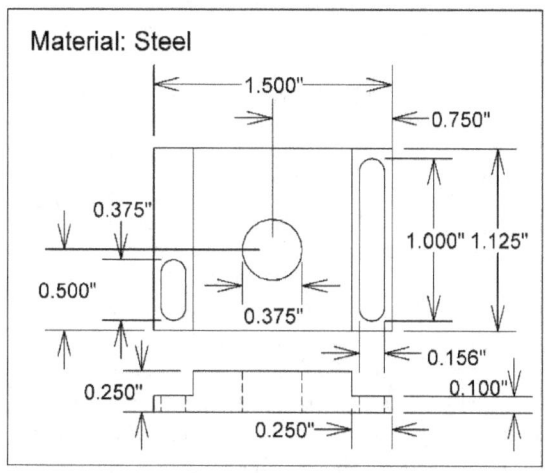

Figure 142 - Split nut carrier plate

The final dimensions chosen for the split nut will depend upon the precise positioning of the leadscrew, and in particular, the distance between the top surface of the leadscrew and the underside of the split nut carrier plate (Figure 142) when the plate is fitted in position. Hence, it is advisable to make and fit the carrier plate first. This is machined from a rectangle of ¼" thick steel, 1 ½" X 1 1/8". The 3/8" diameter hole will take the split nut pivot; the elliptical holes are used to mount the plate under the left-hand front edge of the saddle. The two short edges of the plate are thinned to 0.1" over a ¼" width in order for the mounting screw heads not to foul the split nut. The 3/8" hole is nominally centred ½" from the edge of the plate; however, this dimension should be varied according to the position of the leadscrew.

Measure the horizontal distance from the front face of the saddle to the axis of the leadscrew and use that as the distance for the centre of the hole. The elliptical mounting holes will allow corrections to be made later if necessary. The underside of the saddle casting should be filed or milled to form a flat, horizontal surface to mount the plate on. Spot through the elliptical holes onto the under edges of the casting for three 4BA tapped holes; these will take cheese head or cap head screws to mount the plate in position. Place the holes to allow for fine adjustment of the plate position on final assembly.

Figure 143 shows the split nut itself. Machining commences by facing off both ends of a piece of ¾" square steel stock to a finished length of 2.2", and axially drilling and tapping it 3/8" BSF.

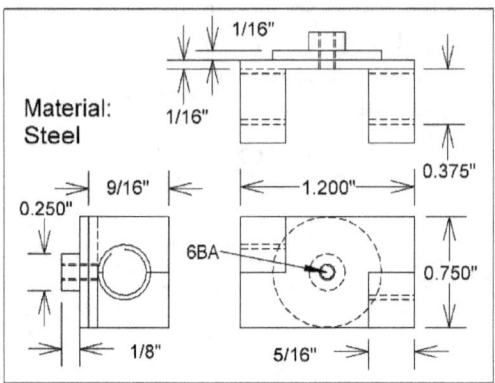

Figure 143 - Split nut

The hole is drilled slightly off centre; the centre of the hole is shown as 5/16" from one face, and 3/8" from each of the two adjacent faces. Setting up and drilling this offset hole is straightforward in the 4-jaw chuck. Still using the 4-jaw, the ¼" diameter spigot is turned on the face that is furthest from the axial hole. The spigot is 1/8" long. A ¾" diameter pad is turned on the same face while the piece is still in the 4-jaw; finally, the spigot is drilled and tapped 6BA. The piece is then milled (or very carefully filed) out to form the two split nuts, as shown in the diagram. Figure 144 shows the finished split nut, assembled onto the mounting plate.

Figure 144 - Split nut and mounting plate

I had the advantage of being able to use a Taig CNC mill to machine the split nut, but it should also be possible using the milling slide in the lathe, or even careful hand filing. There should be 1/16" of material left at the thinnest part of the carrier plate (1/8" thick under the ¾" diameter pad). At this point, if the work has been done accurately so far, the threads should be barely visible on the areas that have been machined out. The dimensions quoted here are based on the assumption that the distance between the carrier plate and the leadscrew is 1/8", as was the case in my prototype. If the measured distance proves to be greater than this, adjust the thickness of the carrier plate either by changing the positioning of the threaded hole (i.e., move the hole further off-centre) or turn down less of the back of the nut (i.e., leave a shorter spigot, and thicker pad). The function of the spigot is merely to locate the nut relative to the shaft that it is mounted on, so can be as little as 1/16" long if need be. If there is much more space than can be accommodated by these adjustments, then try starting from larger stock. Clearly, if you find that there is less than 1/8" clearance below the plate, then it is straightforward to remove more material from the base of the saddle casting to make room.

The split nut is pivoted through the hole in the carrier plate by means of a brass pivot (Figure 145). This is turned down from ½" brass bar; first turn ¼" of its length down to 3/8" diameter. The intent is that this should be a close "running" fit in the hole in the carrier plate. Also, the length of this section should be adjusted to very slightly more than the thickness of the carrier plate, so that the pivot and nut will turn freely, but with minimal play, once assembled. Bore through to clear a 6BA bolt (2.8mm), then counterbore 1¼" diameter to a depth of 1/8" to take the full depth of the spigot. Check that the spigot goes fully home. Part off, reverse and finish to a total length of ½"; counterbore from the ½" diameter end, aiming to leave a shoulder of about 1/16" of metal. Drill for a 3mm (or equivalent size) set screw.

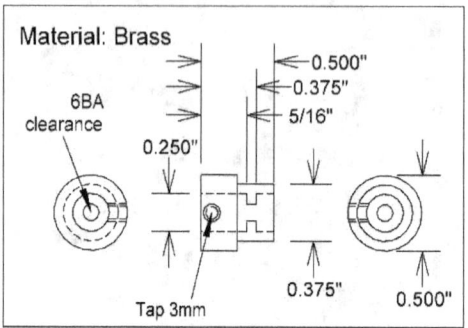

Figure 145 - Split nut pivot

The mounting plate, split nut and its pivot can now be assembled. Push the thin end of the pivot through the hole in the mounting plate, from what will be the upper face when assembled (the non-rebated face). Push the spigot home from the other side. Fit a 6BA cap head screw through the hole in the pivot and tighten the assembly. Check that the split nut pivots freely and that there is minimal play; adjust the pivot length if necessary.

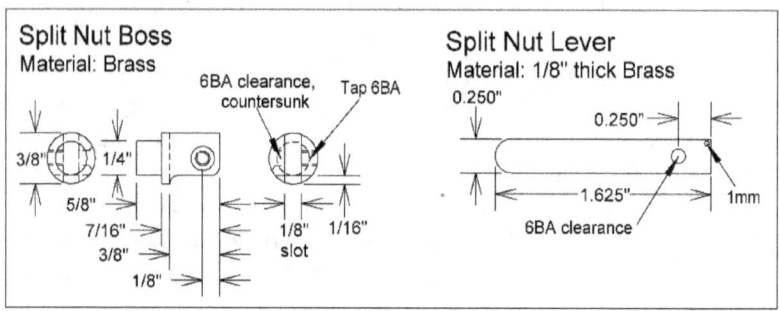

Figure 146 - Split nut boss and lever

The final components are the split nut boss and split nut lever (Figure 146). The latter is simply a length of ¼" X 1/8" brass, with a pivot hole drilled to take a 6BA screw, and a small (1mm diameter) hole. The latter hole will take a small spring, such as the ones that are found in retractable ballpoint pens.

The boss is turned from 3/8" diameter brass. Reduce it to ¼" diameter over a length of 3/16" and part off to a total length of 5/8". Cut a 1/8" wide slot in the fat end, to a depth of 3/8" to take the lever. Cross drill with a 6BA tapping drill, then open out one half to clearance size (2.8mm) and countersink. Tap the other half 6BA.

File off the "back" of the boss, removing about 1/16" of material. The lever can now be fitted, with the short end protruding from the "back" of the boss and pivoted on a 6BA countersunk screw. Make up a short spring, using a ballpoint pen spring or similar, so that the overall length is approximately ½" and there is a loop at either end. Thread one end through the 1mm hole in the operating arm.

The final operations involve cutting a slot in the front face of the saddle to allow the operating arm to protrude through from the back of the saddle. The slot needs to be about ½" long, with two notches about 1/8" deep in the lower edge. The notches locate the operating arm in the engaged and disengaged positions and therefore need to be cut at an angle; the best approach is to cut the slot first, test assemble the lever in position and mark out the positions for the notches. The position of this slot should be such that the arm is horizontal when it is located in either notch; i.e., cut the slot so that the bottom edge of the two notches is approximately 9/16" above the upper face of the carrier plate. Note that to disengage the split nut, the arm does not need to move very far; also, care should be taken to ensure that when in the disengaged position, the split nut does not foul the foot of the lathe when the saddle is near the headstock. I found that the "land" between the two notches needs to be no more than 1/8" wide in order to allow the nut to disengage properly and not to foul the foot, see Figure 141. You will also notice from this photo that much trial & error with this arrangement forced me to cover up earlier (failed) attempts with a small brass plate that carries the final notches; this may prove to be a good solution for others as well!

Final assembly of the split nut involves fitting the arm and boss to the split nut pivot and locating it with the set screw so that, when the operating arm is in the engaged (furthest left) position, the split nut aligns with the axis of the leadscrew. In order to access the set screw, this can only easily be done with the saddle removed from the lathe. It is necessary to provide a means of locating the free end of the operating arm return spring inside the saddle casting. The simplest approach here is to drill a 2.8mm (6BA clearance) hole through the left-hand end of the saddle, about ½" down from the top edge and ½" in from the front. A 6BA screw and nut can then be used to locate and tension the spring. Once in position and tensioned, the spring should be just strong enough to keep the operating arm located in the notches.

Loosen the carrier plate's mounting screws so that the plate can be adjusted for alignment and re-fit the saddle with the operating arm in the disengaged position. Check the operation of the split nut, adjusting the position of the mounting plate as necessary. The aim here is to ensure that, when the nut is in the engaged position, it does not deflect the leadscrew from its normal axis. You may find that a small amount of filing inside the saddle casting is necessary to allow the carrier to be correctly located and for the pivot not to bind against the wall of the casting. Finally, tighten the mounting plate screws.

At this point, it should be possible to engage and disengage the split nut, and to drive the saddle via the leadscrew handwheel.

The "banjo" components

The "banjo" arrangement used in the Sherline kit is slightly unconventional, partly due to the fact that they always use 60 DP gears for the first driver and driven wheels. Two aluminium brackets are provided; the main bracket (Figure 147) mounts on a boss at the left hand end of the leadscrew, and in the normal Sherline set-up, the auxiliary bracket (Figure 148) mounts on the far end of the first.

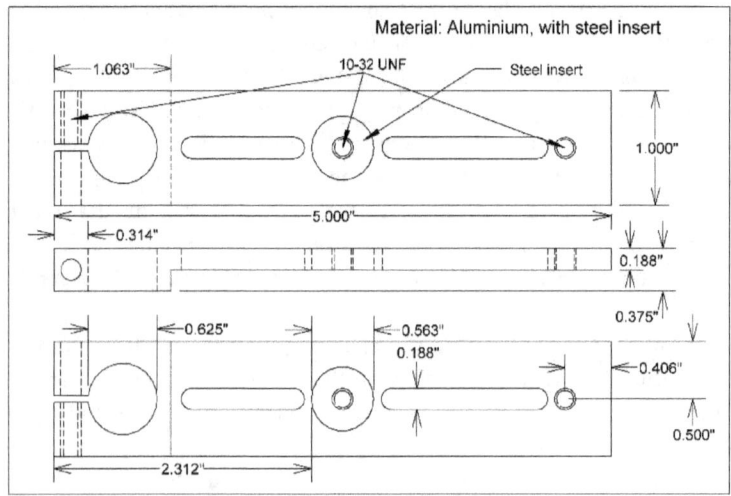

Figure 147 - Banjo main bracket

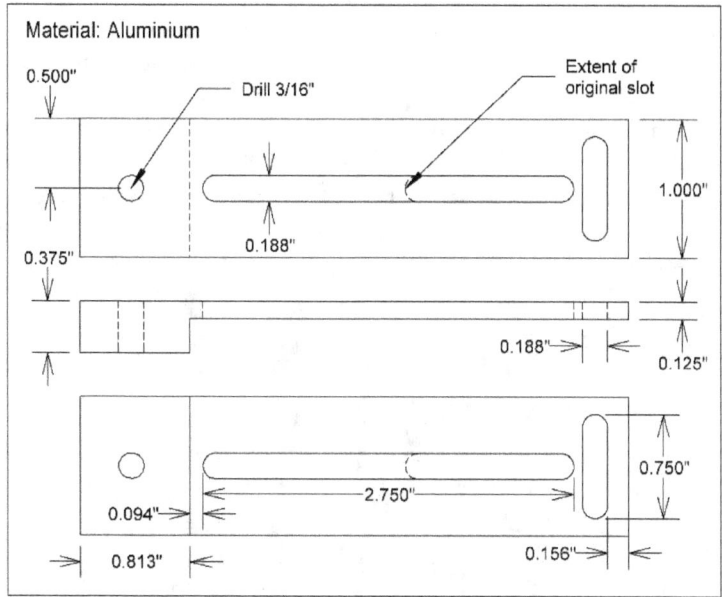

Figure 148 - Banjo auxiliary bracket

Rather than make a purpose-built banjo from scratch, I decided to retain the Sherline parts and modify them to suit the new application. The main bracket mounts on the turned section of the dog clutch body. The diagrams show the modified versions of these components.

Figure 149 shows the components alongside the studs made for use with them. The resultant banjo allows a more (but not completely) conventional arrangement. For many gear trains, the fine feed set-up described below being an example, only the main bracket is required; however, the auxiliary bracket allows a wide range of configurations to be supported. For those who choose not to use the Sherline parts, both of these brackets can be easily constructed from aluminium, brass or steel strip.

---

[19] If making the main bracket from scratch, don't bother with the central threaded hole; simply form one slot by joining the two shown. It might make sense not to bother with the other threaded hole either; simply extend the slot to where the hole is. Also, I would probably not mill the bracket to half thickness; much easier to leave it at 3/8" thick for the whole of its length. However, if using this approach, the long stud base lengths, and the method of fixing the auxiliary arm, may have to be modified.

The main bracket starts out with a 5/16" diameter hole and a 10-32 UNF pinch bolt at one end, a UNF 10-32 threaded hole at the far end, and a 9/16" hole about half way between. The pinch bolt end of the bracket is 3/8" thick; the remainder of its length is milled out at the back of the bracket to half that thickness. The modifications involve cutting the two oval slots shown in the diagram, both 3/16" wide, and filling the central hole with a steel insert, drilled/tapped 10-32 UNF at the centre. The insert is a press fit, assisted with Superglue for good measure[19]. The slot lengths are not critical; the one nearest the pinch bolt end is about 1" long, the other about 1.5" long.

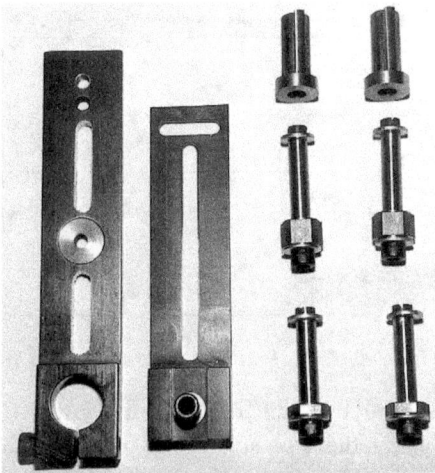

Figure 149 - Banjo brackets and studs

The auxiliary bracket is 4 1/16" long, 3/8" thick for the first 13/16" of its length, and 1/8" thick for the remainder of its length. If making up one of these from scratch, there is little point in retaining the thickened section; use a length of 1/8" thick material. There are two slots in the thin end, arranged in a T, and a 3/16" hole at the thick end to take a 10-32 UNF bolt. The modification to this part involves extending the longitudinal slot towards the thick end; the resultant slot is 2.75" long.

The final components for use with the banjo brackets are the long and short studs, Figure 150. These differ only in the length of their bases; the short studs mount directly on the main banjo bracket and the long studs mount on the auxiliary bracket. The auxiliary bracket is mounted on the "back" of the main bracket (it is mounted using either of the two threaded holes, with a 10-32 socket head bolt through from the back). Hence, the long stud needs a base that is 3/16" longer than the short stud, for the shoulders of these studs to be in the same plane when the banjo is in use.

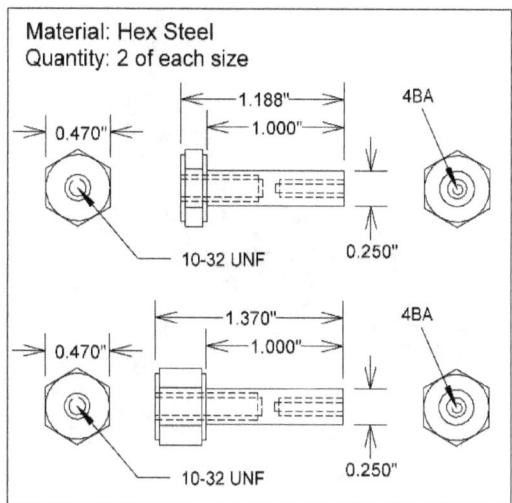

Figure 150 - Short and long studs

Figure 151 shows the banjo assembly, with the auxiliary arm mounted on the main arm, and both types of stud fitted. The studs are turned from hex section steel; nominally 0.470" across flats, but this dimension is unimportant. They consist of a plain-turned section that is nominally 1" long, drilled and tapped 4BA, and a hex section shoulder that is drilled and tapped 10-32 UNF. As before, the hex section is relieved to present a circular section to the bracket on one side, and the bobbin on the other. The 1" length should be just a couple of thou longer than the length of the bobbins, to ensure that the bobbins run freely on the studs when held in place with 4BA retaining screws and washers.

Tony Jeffree

Figure 151 - Banjo assembly

Fine feed set-up

The moment of truth has now arrived. The leadscrew is in place, along with tumbler reverse, dog clutch and split nut, and the banjo components have been constructed or modified, so all that remains is to install some wheels and see what happens.

The leadscrew is 20 TPI, or 50 thou of saddle movement per revolution. Given the size of the lathe, it will generally be the case that its cutting tools will have small tip diameters, and hence, it is highly desirable to employ a really fine feed. In order to achieve this with a small number of gear wheels, I decided to use an overall reduction ratio of 24:1 as the fine feed, achieved by using two 20T driver wheels, one 80T and one 120T driven wheels. This ratio gives a saddle feed of a touch over 2 thou per spindle revolution, which seems to work quite well.

The 20T wheels are again HPC part number G-24-20-PG, the 80T and 120T wheels are part numbers G-24-80-PG and G-24-120-PG respectively. The 120T wheel is a monster, approximately 5" in diameter, and needs to be used as the final driven wheel, held on the clutch input shaft. This is one of the major reasons for allowing suitable "overhang" at the left-hand end of the lathe, as this wheel hangs well below the lathe foot. In order to make use of these wheels some modification is necessary. The 20T wheels come with a bore size of 0.3125", which needs to be opened out to 3/8" using the techniques described earlier. Having done that, all four wheels need a 1/8" wide keyway cut in the bore. The easiest way to do this is to drill a 1/8" hole as close to the bore as possible and to open it out by careful sawing and filing. The wheels should then fit nicely on the various bobbins already constructed.

Figure 152 - Fine feed setup

The fine feed set-up needs just the main banjo arm, fitted with a single short stud and bobbin by means of a 10-32 UNF socket head bolt from the back of the arm. First, fit one 20T wheel to the tumbler output bobbin, retaining it with a 4BA screw & washer. Fit the 80T wheel and second 20T wheel to the stud, using a spacer of a suitable thickness between the two wheels. Suitable spacers can be made up very simply by cutting washers of varying thicknesses (for example, 1/8", 1/4" and 1/2" will prove useful for various configurations), with a 3/8" hole, and a keyway cut as described for the wheel modifications. Fit the Banjo arm to the clutch body and adjust the position of the stud to make it possible to fit the 120T wheel to the clutch input bobbin. It will be necessary to insert more spacers onto the bobbin before the 120T wheel; it should be fitted so that it is flush with the extreme end of the input shaft. Retain the 120T wheel with a 4BA screw & washer. Adjust the stud position so that the 20T wheel meshes nicely with the 120T wheel and tighten up its 10-32 UNF bolt. The banjo can then be rotated up until the 80T wheel meshes nicely with the 20T wheel on the output of the tumbler, and its clamp nut tightened. Figure 152 shows the fine feed installed and ready to go.

At this point, it is advisable to make sure that all bearing surfaces are liberally lubricated, especially those that involve steel-to-steel contact, such as the bobbins and idler wheels.

Disengage the split nuts, set the tumbler in neutral, engage the dog clutch and make sure that all operates smoothly by hand. Engage the tumbler and switch on the lathe. The leadscrew should rotate nicely. Check that the clutch can be operated while the lathe is running; if not, polishing the D sections of the shafts may help.

Try engaging the split nuts. This should not be forced; if all is nicely aligned, the nuts will click into place quite easily with a little rotation of the leadscrew, and the handle should be kept seated in its slot by the spring tension. If all is well, you are ready to use powered fine feed on your Taig lathe for the first time.

Obviously, the choice of gears described here is only one possible fine feed configuration; the use of a second stud opens up the possibility of using smaller wheels, albeit having to use 6 rather than 4. One of the considerations, however, was that the 120T wheel seems like a useful wheel to have around in order to use the change wheels as the basis of dividing operations in the lathe. Another observation here is that if these wheels are to be used only for fine feed (i.e., if you do not intend to inter-mix them with the Sherline wheels for any operations), then there is no particular need to stick to 24DP. The essential requirement is that the bore, keyway and thickness of the wheels allow them to be fitted to the bobbins. Equally, there is no particularly good reason why a fine feed shouldn't be configured using suitably sized pulleys and belts rather than gears, again as long as they can be fitted to the bobbins. In fact, pulleys and belts are in some respects more desirable, as they produce a lot less noise!

With the fine feed set-up described, it is quite acceptable to engage/disengage the split nut and the dog clutch while the leadscrew is rotating; however, never try operating the tumbler reverse while the lathe is running.

Screw cutting set-up

This is achieved in a similar manner to the fine feed set-up. For many, if not all, of the screw pitches that you will want to use, the 24DP wheels in the Sherline kit can be used and can be combined with the 24DP wheels used for the fine feed configuration. The banjo arrangement allows a wide range of configurations; if necessary, further bobbins can be made up in order to allow more than 2 intermediate studs to be used if this proves to be beneficial.

The 60DP wheels provided in the Sherline kit are one 50T, two 100T, and one 127T wheel. These wheels can obviously only be used as a matched driver/driven pair, as they cannot mesh with the other wheels. Clearly, using both 100T wheels is not very interesting as it is easier to omit them altogether, and the 50T/100T combinations provide ratios of 1:2 or 2:1 that can be generated using a 20T and a 40T wheel (or 40T and 80T) from the 24DP set. Hence, it will almost certainly only be necessary to use the 60DP wheels if you wish to make use of the lathe to cut metric threads, using the 127T wheel in combination with a 50T or 100T wheel. In order to make use of these wheels it is necessary to adapt them to fit the bobbins. As supplied, they have a 9/16" bore and a keyway that is about 7/32" wide. These need to be reduced to give a bore size of 3/8" with a 1/8" wide keyway to match the 24DP wheels. The simplest approach here is to turn a length of steel tube, with a 3/8" bore and an OD that will give a press-fit into the 9/16" bore. The keyway can then be milled or hack-sawed straight through the wall of the tube, leaving it with a "C" shaped section. This can be cut into lengths to form adapters for the 60DP wheels, pressed and/or Superglued in place. I chose to cut these to 5/16" in length, to match the thickness of the HPC gears, but this is not essential. Figure 153 shows three modified Sherline wheels; 50T, 100T and 127T.

The table of change wheel combinations given in the Sherline kit is a good starting point for use with this leadscrew. One obvious point is that there is no need to bother with the left hand and right-hand thread variants, as the tumbler allows a single set-up to cut either handedness.

Figure 153 - Modified 60DP wheels

The possible change wheel combinations include threads as coarse as 10TPI; a little thought, given that the leadscrew is 20TPI, leads you to the conclusion that, when cutting a 10TPI thread, the leadscrew will be rotating twice as fast as the lathe spindle. In the case of my own lathe, the slowest spindle speed attainable is around 435 RPM; hence, when cutting 10TPI under power at the low speed setting, the leadscrew would be doing a cool 900 RPM! I have actually tried this; interestingly enough, the experiment was one of the reasons why the tumbler plate is now so substantial, as an early prototype version of the tumbler plate operating arm (which can be seen in Figure 114) buckled and broke under the forces exerted on it by the gear train! This is one of the reasons why Sherline include a substantial, 4" diameter, handwheel in their kit, which is used to hand-drive the spindle at more leisurely speeds. Another reason is that the Sherline kit uses the 60DP wheels in all setups; I doubt whether these are really up to the job of operating under power. Hence, when cutting coarse threads, and when using the metric conversion wheel, running under power is not a great idea.

Similarly, attempting to engage the dog clutch under power with a coarse screw cutting set-up is not a great idea; the shock load on the gear train when the dog engages is a little disconcerting! However, disengaging the clutch under power is not a problem.

A word on safety

It cannot be emphasized too strongly that exposed gear trains have considerable ability to attract and consume loose objects, such as ties, hair, fingers and so on. With the fine feed configuration described here, the 24:1 reduction in rotational speed is accompanied by a corresponding 24:1 increase in torque. Needless to say, getting anything that contains nerve endings trapped in these gears is going to be, at best, a very uncomfortable experience, and at worst, may be extremely damaging to your health. Hence, my advice to anyone that builds and uses this design is to cover the change wheels during operation at all times. I have not offered a cover design here, as mounting arrangements will inevitably differ according to baseboard layouts and so on; however, it should be a trivial job to fabricate a suitable cover from sheet metal – e.g., thin plywood, or acrylic sheet - in such a manner that it can be hinged out of the way for adjustments to the gear train and secured in position while in operation.

Annex A
Suppliers and other resources

This section includes supplier information, plus pointers to useful resources for the desk top machinist that can be found on the Internet. As suppliers come and go, this list will inevitably become out of date over time; these days, a search engine may be of more help!

Taig Tools

Taig Tools, 12419 E. Nightingale Lane, Chandler AZ 85286, USA.
Phone - +1 480 895 6978
http://www.taigtools.com

Taig dealers

Taig Tools keep a list of Taig dealers worldwide on their website, here:
http://www.taigtools.com/dealers

Tool suppliers

The following suppliers offer indexable carbide insert tooling with 1/4" or 6mm shanks:

J.B. Cutting Tools Ltd, 19 Princess Road, Dronfield, Sheffield, S18 2LX, U.K.
Tel: +44-1246-418110
http://www.jbcuttingtools.com

Greenwood Tools, 2a Middlefield Road, Bromsgrove, Worcs, B60 2PW, U.K.
Tel: +44-1527-877576
https://www.greenwood-tools.co.uk/

Chronos, Unit 2 Southfields Road, Dunstable, LU6 3EJ, U.K.
Tel: +44-1582-471900
http://www.chronos.ltd.uk/

Other suppliers

HPC Gears, Unit 14, Foxwood Industrial Park, Chesterfield, S41 9RN, UK.
Tel: 01246-268080
http://www.hpcgears.com/

Sherline tools and accessories can be obtained from:
Sherline Products Inc., 3235 Executive Ridge, Vista, CA 92083-8527, USA
Tel: +1 800-541-0735 (Toll free for USA/Canada) or +1 760-727-5857
For distributors in other countries, visit the Sherline website at:
http://www.sherline.com.

RS Components, P.O. Box 33, Corby, Northants, NN17 9EL, U.K.
Tel: 01536 204555
https://uk.rs-online.com

Internet resources

http://www.cartertools.com
Nick Carter, a Taig dealer from oregon, runs this website; it contains a wealth of information on the Taig machines, including many practical and constructional articles and photos contributed by other Taig enthusiasts.

http://www.lathes.co.uk/index.html

THE TAIG/PEATOL LATHE

Tony Griffiths, a used machinery dealer in the UK, has constructed a phenomenal archive of information on all kinds of lathes, milling machines, and other workshop equipment.

http://groups.yahoo.com/group/taigtools

Yahoo! runs a large number of eGroups on various topics; these allow members to post Email messages and files to the group. This group is concerned with the Taig lathe and mill and has a very enthusiastic following from all corners of the globe. Lots of useful advice is available for the beginner and the more advanced machinist too; ask a question of the group and you will get a good number of sensible answers.

http://www.jeffree.co.uk/modelengineering.html

My website concerned with model engineering topics and articles. Some of the material on this site (particularly photos and diagrams) may look familiar, as it has been re-used in writing this book

Annex B
Lathe specifications

General Specifications

- Overall working accuracy: 0.0005"
- Maximum bearing runout: 0.0004"
- Headstock normality to bed: 0.0004" max error
- Cross slide normality to bed: 0.0004" max error
- Max taper bed dovetail over pins 0.0001"
- All machine dovetails: 45 degrees
- Bed width: 2.875"
- Cross slide dial graduations: 0.001"
- Cross slide screw: 1/4", 20 TPI
  (Backlash adjustment provided—jam nut arrangement)
- Carriage travel per revolution of the handwheel: 0.500"
- Max spindle speed recommended: 7000 RPM
- Motor requirement: 1/8 to 1/4 HP
- Pulley type: standard 5/8" bore, multi-step, V-belt
- Length of headstock on ways: 2.875"
- Length of carriage on ways: 3"
- Width of cross slide on carriage: 2"
- Tool post may be adjusted for angle cutting, chamfering, boring etc.

THE TAIG/PEATOL LATHE

Capacity

— Swing over bed: 4.5"
  (Max turning diameter 4.5")
— Swing over cross slide: 2.875"
— Overall length of bed: 15"
— Overall length of lathe: 16.5"
— Tool bit size: 1/4" shank
— Distance between centres: 9.75"
  (Tailstock optional)
— Carriage travel: 9"
— Cross-slide travel: 1.75"

Spindle

—Sealed precision bearings: 40mm (1.5748") OD, 17mm (0.6692") ID
—Spindle nose thread: 3/4"-16 (3/4" SAE)
—Spindle through hole: 0.343"
—Spindle ID taper: 15 degrees (30 degrees included angle)
—Max. collet diameter: 9/32"
—Pulley size: 5/8" bore

www.ingramcontent.com/pod-product-compliance
Lightning Source LLC
Chambersburg PA
CBHW070631220526
45466CB00001B/151